H. Sancouci
-privat-

in situ HYBRIDIZATION in LIGHT MICROSCOPY

Methods in Visualization

Series Editor: Gérard Morel

In Situ Hybridization in Light Microscopy
Gérard Morel and Annie Cavalier

Visualization of Receptors: In Situ Applications of Radioligand Binding
Emmanuel Moyse and Slavica M. Krantic

Genome Visualization by Classic Methods in Light Microscopy
Jean-Marie Exbrayat

Imaging of Nucleic Acids and Quantification in Phototonic Microscopy
Xavier Ronot and Yves Usson

In Situ Hybridization in Electron Microscopy
Gérard Morel and Annie Cavalier

in situ HYBRIDIZATION in LIGHT MICROSCOPY

**Gérard Morel, Ph.D., D.Sc.
Annie Cavalier, Eng.**

CRC Press
Boca Raton   London   New York   Washington, D.C.

Library of Congress Cataloging-in-Publication Data

Morel, Gérard.
  [Hybridation in situ en microscopie photonique. English]
  In situ hybridization in light microscopy / Gérard Morel, Annie Cavalier.
      p.   cm. — (Methods in visualization)
  Includes index.
  ISBN 0-8493-0703-1 (alk. paper)
  1. In situ hybridization. 2. Microscopy. I. Cavalier, Annie. II. Title. III. Series.

QH452.8 .M67 2000
572.8'4—dc21   00-057199
                                                                                        CIP

This book contains information obtained from authentic and highly regarded sources. Reprinted material is quoted with permission, and sources are indicated. A wide variety of references are listed. Reasonable efforts have been made to publish reliable data and information, but the author and the publisher cannot assume responsibility for the validity of all materials or for the consequences of their use.

Neither this book nor any part may be reproduced or transmitted in any form or by any means, electronic or mechanical, including photocopying, microfilming, and recording, or by any information storage or retrieval system, without prior permission in writing from the publisher.

The consent of CRC Press LLC does not extend to copying for general distribution, for promotion, for creating new works, or for resale. Specific permission must be obtained in writing from CRC Press LLC for such copying.

Direct all inquiries to CRC Press LLC, 2000 N.W. Corporate Blvd., Boca Raton, Florida 33431.

Trademark Notice: Product or corporate names may be trademarks or registered trademarks, and are used only for identification and explanation, without intent to infringe.

© 2001 by CRC Press LLC

No claim to original U.S. Government works
International Standard Book Number 0-8493-0703-1
Library of Congress Card Number 00-057199
Printed in the United States of America 1 2 3 4 5 6 7 8 9 0
Printed on acid-free paper

SERIES PREFACE

Visualizing molecules inside organisms, tissues, or cells continues to be an exciting challenge for cell biologists. With new discoveries in physics, chemistry, immunology, pharmacology, molecular biology, analytical methods, etc., limits and possibilities are expanded, not only for older visualizing methods (photonic and electronic microscopy), but also for more recent methods (confocal and scanning tunneling microscopy). These visualization techniques have gained so much in specificity and sensitivity that many researchers are considering expansion from in-tube to *in situ* experiments. The application potentials are expanding not only in pathology applications but also in more restricted applications such as tridimensional structural analysis or functional genomics.

This series addresses the need for information in this field by presenting theoretical and technical information on a wide variety of related subjects: *in situ* techniques, visualization of structures, localization and interaction of molecules, functional dynamism *in vitro* or *in vivo*.

The tasks involved in developing these methods often deter researchers and students from using them. To overcome this, the techniques are presented with supporting materials such as governing principles, sample preparation, data analysis, and carefully selected protocols. Additionally, at every step we insert guidelines, comments, and pointers on ways to increase sensitivity and specificity, as well as to reduce background noise. Consistent throughout this series is an original two-column presentation with conceptual schematics, synthesizing tables, and useful comments that help the user to quickly locate protocols and identify limits of specific protocols within the parameter being investigated.

The titles in this series are written by experts who provide to both newcomers and seasoned researchers a theoretical and practical approach to cellular biology and empower them with tools to develop or optimize protocols and to visualize their results. The series is useful to the experienced histologist as well as to the student confronting identification or analytical expression problems. It provides technical clues that could only be available through long-time research experience.

Gérard Morel, Ph.D.
Series Editor

GENERAL INTRODUCTION

Molecular biology and biotechnology allow the study of the genome and its expression through the knowledge acquired on nucleic acids, which are the basis of genetics. Thus, the study of transcription mechanisms in normal or pathological cells is fundamental for understanding normal and tumoral differentiation. It is now possible to visualize a given sequence in the total genome, as well as to reveal the presence of a virus in a single cell within normal tissue. It is therefore now essential in both clinical diagnosis and pharmaceutical analysis to use such techniques in studying the complete genome and its transcription.

Furthermore, this visualization allows an integration of this detection with the basic physiological and pathological mechanisms found in the whole organ.

Although many visualization techniques are now published, there are few critical analyses of these methods. Therefore we have attempted to present an analytical approach to all of the methods involved in nucleic acid visualization, thus bringing out all the variations and possibilities and allowing readers to develop their own methodology for the problem concerned.

The aim of this book is to present an analysis on the different stages of *in situ* hybridization by light microscopy, thus merging both theory and practice.

The presentation in two columns enables the theory and the practical techniques, illustrated by numerous diagrams, to be viewed face-to-face with commentaries and references. Several tables suggest potential criteria of choice and decisions to the reader.

The work is divided into eight chapters, each presenting one stage of the method: tissue preparation; pretreatment; hybridization itself; detection of the hybrid; and several methods of counterstaining and observation. The different types of probes and the techniques for labeling them are analyzed. A chapter is devoted to basic protocols that can guide the newcomer through the correct methodology. Another chapter analyzes the controls, the problems of sensitivity, and the relationship between specific and nonspecific labeling. The method of preparation of different reagents and solutions, their sterilization, and their storage are described in the appendices, which are followed by a glossary of principal terms used. Finally, an index allows the reader to locate subjects at a glance.

ACKNOWLEDGMENTS

The authors would particularly like to thank Professor Tomás Garcia-Caballero (Department of Morphological Sciences, School of Medicine, Santiago de Compostela University, Santiago de Compostela, Spain) and Doctor Christopher Perrett (Department of Obstetrics and Gynecology, Royal Free and University College Medical School, University College London) for their participation in the preparation and proofreading of the manuscript, as well as their colleagues who contributed to the successful completion of this project. This work is dedicated to the memory of Professor Jean Gouranton, Ph.D. (1934 to 1999), an exemplary friend and scientist.

The work was carried out in the framework of the European "Leonardo da Vinci" project (Grant F/96/2/0958/PI/II.I.1.c/FPC), in association with Claude Bernard-Lyon 1 University (*http://brise.ujf-grenoble.fr/LEONARDO*).

A number of the illustrations were kindly supplied by Pr. Tomás Garcia-Caballero.

English translation: John Doherty

THE AUTHORS

Gérard Morel, Ph.D., D.Sc., is Research Director at the National Center of Scientific Research (CNRS), at the University Claude Bernard-Lyon 1, France. Dr. Morel obtained his M.S. and Ph. D. degrees in 1973 and 1976, respectively, from the Department of Physiology of Claude Bernard University-Lyon1. He was appointed Assistant in Histology at Claude Bernard University in 1974 and became Doctor of Science in 1980. He was appointed by CNRS in 1981 and became Research Director in 1989. Dr. Morel is a member of the American Endocrine Society, International Society of Neuroendocrinology, Society of Neuroscience, American Society of Cell Biology, Société Française des Microscopies, Société de Biologie Cellulaire de France, and the Société de Neuroendocrinologie Expérimentale. He has been the recipient of research grants from the European Community, INSERM (National Institute of Health and Medical Research), La Ligue contre le Cancer, l'ARC (Association de Recherche contre le Cancer), Claude Bernard University, and private industry. Current major research interests of Dr. Morel include internalization (in particular, nuclear receptors for peptides), the regulation of gene expression, and paracrine interactions (low gene expression level of ligand in target tissue).

Annie Cavalier is an engineer in biology at the CNRS at the University of Rennes 1, France. She was appointed a technical assistant in cytology in 1969. To improve her knowledge in the fields of cell biology and molecular biology, she has regularly participated in advanced courses. Annie Cavalier was appointed engineer at the CNRS (1993) and was graduated from the University Claude Bernard-Lyon 1. She is a member of the Société Française des Microscopies and was actively involved in the organization of the 13th International Congress of Electron Microscopy (Paris, July 1994), and the first Congress of the Société Française des Microscopies (Rennes, June 1996). She has contributed to many scientific papers, chapters, and two books. Her current major scientific interest concerns the structural and functional studies of water channels (aquaporins) and glycerol facilitators. Because of her unique technical skill, she is also involved in many collaborations with other scientific groups.

CONTENTS

General Introduction	vii
Abbreviations	xiii
Chapter 1 - Probes	1
Chapter 2 - Tissue Preparation	61
Chapter 3 - Pretreatments	89
Chapter 4 - Hybridization	107
Chapter 5 - Revelation of Hybrids	139
Chapter 6 - Counterstaining and Modes of Observation	187
Chapter 7 - Controls and Problems	207
Chapter 8 - Typical Protocols	221
Examples of Observations	251
Appendices	261
A - Equipment	267
B - Reagents	273
Glossary	303
Index	319

ABBREVIATIONS

AEC	↪ 3-amino-9-ethylcarbazole
ATP	↪ adenosine triphosphate
BCIP	↪ 5-bromo-4-chloro-3-indolyl phosphate
bp	↪ base pair
Bq	↪ becquerel
cDNA	↪ complementary deoxyribonucleic acid
Ci	↪ curie
CTP	↪ cytosine triphosphate
cpm	↪ count per minute
DNA	↪ deoxyribonucleic acid
DAB	↪ 3' diaminobenzidine tetrahydrochloride
dATP	↪ deoxy-adenosine triphosphate
dCTP	↪ deoxy-cytosine triphosphate
DEPC	↪ diethyl-pyrocarbonate
dGDP	↪ deoxyguanosine-5'-diphosphate
dGMP	↪ deoxyguanosine-5'-monophosphate
dGTP	↪ deooxyguanosine 5'-triphosphate
DNase	↪ deoxyribonuclease
dNTP	↪ deoxynucleoside triphosphate
DTT	↪ dithiotreitol
dUTP	↪ deoxyuridine-5'-triphosphate
EDTA	↪ ethylene diamine tetra-acetic acid
Fab	↪ immunoglobulin fragment obtained by proteolysis (papain)
F(ab')$_2$	↪ immunoglobulin fragment obtained by proteolysis (pepsin)
Fc	↪ immunoglobulin fragment obtained by proteolysis (papain)
FITC	↪ fluorescein isothiocyanate
GTP	↪ guanosine triphosphate
HRP	↪ horseradish peroxidase
Ig	↪ immunoglobulin
IgG	↪ immunoglobulin G
Kb	↪ kilobase
kBq	↪ kilobecquerel
kDa	↪ kilodalton
MW	↪ molecular weight
mRNA	↪ messenger ribonucleic acid
NBT	↪ nitroblue tetrazolium
NTP	↪ nucleoside triphosphate
OD	↪ optical density
oligo(dT)	↪ oligo-deoxythymidine
PBS	↪ phosphate buffer saline
PCR	↪ polymerase chain reaction
PF	↪ paraformaldehyde

Abbreviations

RNA	↪ ribonucleic acid
RNase	↪ ribonuclease
rRNA	↪ ribosomic ribonucleic acid
RT	↪ room temperature
RT – PCR	↪ reverse transcription – polymerase chain reaction
SSC	↪ standard saline citrate
T_m	↪ melting temperature
tRNA	↪ transfer ribonucleic acid
T_w	↪ washing temperature
U	↪ unit (enzymatic activity)
UDP	↪ uridine-5'-diphosphate
UTP	↪ uridine-5'-triphosphate
$^v/_v$	↪ volume / volume
$^w/_v$	↪ weight / volume

Chapter 1

Probes

Contents

Definition			7
1.1	Types of Probes		8
	1.1.1	Double-Stranded DNA Probes	8
	1.1.2	Single-Stranded DNA Probes	8
	1.1.3	Oligonucleotide Probes	9
	1.1.4	RNA Probes	11
	1.1.5	Criteria of Choice	12
1.2	Probe Labels		12
	1.2.1	Radioactive Labels	13
		1.2.1.1 Types	13
		1.2.1.1.1 ^{35}S	13
		1.2.1.1.2 ^{33}P	14
		1.2.1.1.3 ^{32}P	14
		1.2.1.1.4 ^{3}H	14
		1.2.1.2 Position of the Label	15
		1.2.1.3 Advantages	16
		1.2.1.4 Disadvantages	16
	1.2.2	Antigenic Labels	17
		1.2.2.1 Types	17
		1.2.2.1.1 Biotin	17
		1.2.2.1.2 Digoxigenin	18
		1.2.2.1.3 Fluorescein	19
		1.2.2.1.4 Photoactivated Molecules	19
		1.2.2.2 Advantages	20
		1.2.2.3 Disadvantages	20
	1.2.3	Criteria of Choice	21
1.3	Labeling Techniques		22
	1.3.1	Random Priming	22
		1.3.1.1 Principle	22
		1.3.1.2 Summary of the Different Steps	24
		1.3.1.3 Equipment/Reagents/Solutions	25
		1.3.1.4 Protocol for Radioactive Probes	26
		1.3.1.5 Protocol for Antigenic Probes	27
	1.3.2	Nick Translation	28
		1.3.2.1 Principle	28
		1.3.2.2 Summary of the Different Steps	29
		1.3.2.3 Equipment/Reagents/Solutions	29
		1.3.2.4 Protocol for Radioactive Probes	30
		1.3.2.5 Protocol for Antigenic Probes	31
	1.3.3	Amplification by PCR	32
		1.3.3.1 Principle	32
		1.3.3.2 Summary of the Different Steps	33
		1.3.3.3 Equipment/Reagents/Solutions	34
		1.3.3.4 Reaction Mixture for Radioactive Labeling	34
		1.3.3.5 Reaction Mixture for Antigenic Labeling	35

		1.3.3.6	PCR Protocol	35
	1.3.4	Amplification by Asymmetric PCR		36
		1.3.4.1	Principle	36
		1.3.4.2	Summary of the Different Steps	36
		1.3.4.3	Equipment/Reagents/Solutions	37
		1.3.4.4	Reaction Mixture for Radioactive Labeling	38
		1.3.4.5	Reaction Mixture for Antigenic Labeling	39
		1.3.4.6	Protocol for Asymmetric PCR	39
	1.3.5	3' Extension		40
		1.3.5.1	Principle	40
		1.3.5.2	Summary of the Different Steps	41
		1.3.5.3	Equipment/Reagents/Solutions	41
		1.3.5.4	Protocol for Radioactive Probes	42
		1.3.5.5	Protocol for Antigenic Probes	43
	1.3.6	5' Extension		43
		1.3.6.1	Principle	43
		1.3.6.2	Summary of the Different Steps	44
		1.3.6.3	Equipment/Reagents/Solutions	44
		1.3.6.4	Protocol for Radioactive Probes	45
	1.3.7	Addition		45
		1.3.7.1	Principle	45
		1.3.7.2	Summary of the Different Steps	46
		1.3.7.3	Equipment/Reagents/Solutions	46
		1.3.7.4	Protocol	47
	1.3.8	*In Vitro* Transcription		48
		1.3.8.1	Principle	48
		1.3.8.2	Summary of the Different Steps	49
		1.3.8.3	Equipment/Reagents/Solutions	50
		1.3.8.4	Protocol for Radioactive Probes	51
		1.3.8.5	Protocol for Antigenic Probes	52
1.4	Precipitation Techniques			53
	1.4.1	Ethanol Precipitation		53
		1.4.1.1	Principle	53
		1.4.1.2	Equipment/Reagents/Solutions	54
		1.4.1.3	Protocol	54
	1.4.2	Passage through a Column		55
		1.4.2.1	Principle	55
		1.4.2.2	Equipment/Solutions	55
		1.4.2.3	Protocol	55
	1.4.3	Drying		56
1.5	Controls for Labeling			56
	1.5.1	Radioactive Probes		56
		1.5.1.1	Calculating the Incorporated Radioactivity	56
		1.5.1.2	Calculation of Specific Activity	56
		1.5.1.3	Verification of Labeling on an Analytical Gel	56
	1.5.2	Antigenic Probes		57
		1.5.2.1	Controls	57
		1.5.2.2	Biotin and Digoxigenin Detection	57
		1.5.2.3	Visualization of Fluorescein	58

		1.5.2.4	Disadvantages	58
1.6	Storage / Use			58
	1.6.1	Storage		58
	1.6.2	Use		59

DEFINITION

A probe is a labeled nucleic acid (DNA or RNA) whose nucleotide sequence is complementary to the target nucleic acid (Figure 1.1).

The probe structure is of one of the following types:
- Double-stranded DNA
- Single-stranded DNA
- Synthetic DNA
- RNA

and the target nucleic acid sequences include:
- DNA in interphase nuclei
- Chromosomal DNA
- Ribosomal RNA
- Transfer RNA
- Messenger RNA
- Viral RNA

⇝ The orientation of the probe nucleotide sequence must be anti-sense to that of the target.
⇝ The label must be chosen according to the visualization method used (*see* Chapter 5).

⇝ Generally cDNA
⇝ Generally obtained by PCR
⇝ Oligonucleotide
⇝ Generally cRNA

⇝ Principle of probe hybridization against target nucleic acid:
- The target is oriented 3' → 5'
- The probe is oriented 5' → 3',
i.e., anti-sense

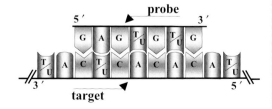

Figure 1.1 Definition of a probe.

1.1 TYPES OF PROBES

Four types of probes can be defined, according to the type of nucleic acid (DNA or RNA) and its method of generation.

↬ Probe (labeled polynucleotides)

1.1.1 Double-Stranded DNA Probes

These probes are double-stranded fragments:
- Of genomic DNA corresponding to a specific sequence, which has previously been cloned

↬ These sequences can contain regions that are not expressed in the cell (introns). They can be used in viral or genome research.

- Complementary to RNA

↬ In this case, it is generally a matter of obtaining DNA complementary to the sequence of messenger RNA (complementary DNA: cDNA) whose sequence is published in the literature, or in a data bank.

- Amplified by PCR

↬ From the given sequence, it is possible to amplify (in an exponential fashion) a particular sequence that will serve as the probe.

The DNA is inserted into a plasmid (e.g., pBR322, pGEM-13).

↬ The plasmids must be amplified using bacteria; the DNA is then extracted and purified on a gel before being cut with restriction enzymes to isolate the insert.

❑ *Advantages*
- The probes are very stable.
- Amplification uses a bacterial culture.
- The labeling protocol is well established.
- Labeling is done at the same time as PCR synthesis.

↬ Stability of double-stranded DNA probes
↬ Allows the storage of a renewable stock
↬ Using random priming or nick translation
↬ Limited by the PCR possibilities

❑ *Disadvantages*
- It is necessary to obtain the cloning plasmid from the originator.
- The probe must be amplified.

↬ Sometimes difficult

↬ Necessary to have reagents and equipment for amplification

- Rehybridization of double-stranded probes is necessary.

↬ Necessary to denature before use

1.1.2 Single-Stranded DNA Probes

These probes are:
- Fragments amplified by PCR

↬ From a given sequence, it is possible to amplify (in an arithmetical way) a sequence that will serve as the probe.

- Single-stranded DNA fragments inserted into a bacteriophage vector (e.g., M13 and its derivatives).
❏ *Advantages*
- The probes are very stable.
- They are available in large amounts.
- There is no need to denature.
- There is no hybridization between probes.

- They are labeled at the same time as PCR synthesis.
❏ *Disadvantages*
- The cloning plasmid must be obtained from the originator.
- The probe must be amplified.

- There is relatively weak specific activity.

⇝ It is necessary to amplify the plasmids with the aid of bacteria.

⇝ DNA probes
⇝ After amplification
⇝ Except in the case of palindromic sequences
⇝ Possibility of palindromic sequences that lead to intrachain hybridization
⇝ Limited by PCR yield

⇝ Sometimes difficult

⇝ Necessary to have reagents and equipment for amplification
⇝ Labeling techniques limited to 5' end labeling or 3' extension

1.1.3 Oligonucleotide Probes

These probes are of single-stranded synthesized DNA whose sequence is complementary to that of the target nucleic acid.
The criteria for construction are:
- A complementary sequence to the target nucleic acid sequence

- A sequence comprising a coding region

- A sequence of between 20 and 50 nucleotides

- A G + C content of ≤ 50 to 55%

- An absence of palindromic sequences

- The sequence compared with that in a gene bank

⇝ The sequences are published in the literature or in the various data banks.
⇝ They are accessible commercially.
⇝ The choice of sequence is essential.
⇝ The published cDNA generally corresponds to the mRNA converted into cDNA. The probe sequence must be the complementary and antisense sequence to this cDNA (Figure 1.2).
⇝ This is especially necessary for mRNA research.
⇝ An oligonucleotide is generally specific for a target sequence of more than 18 nucleotides.
⇝ Above 60% G + C, nonspecific interactions have been found.
⇝ These lead to intrachain or interchain loops, which inhibit hybridization.
⇝ This avoids the synthesis of unusual or repetitive sequences. This work, assisted by a computer, is never exhaustive, but limits this type of problem.

1 = Example of oligonucleotide probe construction specific for rat growth hormone (rGH) from a cDNA sequence.

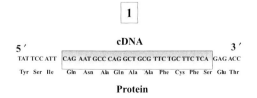

Probes

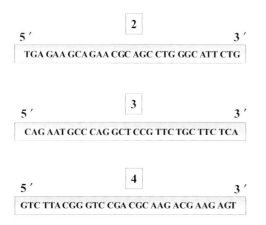

2 = **Anti-sense probe** — The complementary sequence is anti-sense to the cDNA.

3 = **Sense probe** — Negative control probe.

4 = **Non-sense probe** — Negative control probe.

Figure 1.2 Determination of the probe sequence.

❑ *Advantages*
- Very accurate determination of a specific sequence

- Very stable probe
- Available in large amounts

- Well-established procedure for labeling

- Synthesis of labeled probe

- Synthesis of sense probe
- Synthesis of non sense probe

- Quantification of the signal

- Simultaneous use of multiple oligonucleotides

❑ *Disadvantages*
- Difficulty determining the *in situ* hybridization and washing temperatures
- Simultaneous use of multiple oligonucleotides

- Generally labeled by 3' extension

- Efficiency of 3' extension limited in the case of antigenic nucleotides

↝ In the case of strong homologies between different nucleic acids coding for related proteins, the choice of a limited and very specific sequence is the only possibility.
↝ DNA probes are stable.
↝ Probes are easy to obtain commercially in large quantities.
↝ 3' extension is the most widely used method for *in situ* hybridization.
↝ The incorporation of a labeled nucleotide can be achieved in the course of synthesis (in the case of antigenic nucleotides, the efficiency of labeling is known).
↝ Their synthesis is easy to achieve, and produces excellent negative controls: sense and non-sense probes do not allow hybridization, and produce no signal (*see* Figure 1.2).
↝ An oligonucleotide hybridizes to only one target nucleic acid.
↝ Signal is amplified.

↝ Parameters are to be determined experimentally.
↝ The hybridization and washing conditions must be compatible.
↝ The precise control of extension length is very difficult to achieve.
↝ The addition of unlabeled nucleotides can help.

1.1.4 RNA Probes

These probes are of single-stranded RNA (riboprobes) resulting from the transcription of a DNA insert subcloned into a plasmid (Figure 1.3).

The riboprobes are obtained by *in vitro* transcription from one strand of the insert (either strand can be transcribed into RNA). The plasmid is first linearized using a restriction enzyme site downstream from the insert in relation to the polymerase used.

↬ This DNA insert (double or single stranded) must be situated in a circular plasmid containing the binding sites for RNA polymerase (promoters), allowing the transcription of this DNA insert into RNA.
↬ If the transcription takes place from the opposite and complementary strand to the mRNA, an anti-sense probe is obtained.

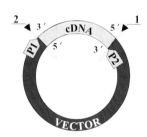

↬ The promoters for RNA polymerases are generally SP6, T3, or T7.
↬ Cleavages (1) and (2) are obtained using restriction enzymes characteristic for the vector or plasmid.
P1 = Promoter 1
P2 = Promoter 2

Figure 1.3 Insert in a plasmid.

❏ *Advantages*
- Total synthesis of the probe

- Labeling during synthesis

- High specific activity

- Very stable hybrids

- Sense probe

❏ *Disadvantages*
- Difficult preparation

- Difficult storage

↬ They must be cut into fragments of 200 to 300 nucleotides
↬ Synthesis of the RNA by *in vitro* transcription allows the incorporation of labeled nucleotides. The whole probe is labeled.
↬ This depends on the number of labeled nucleotides used, and the ratio of labeled to unlabeled nucleotides.
↬ The stability of the hybrids is in increasing order: DNA–DNA, DNA–RNA, RNA–RNA.
↬ If the transcription is carried out using the strand coding for the mRNA, the probe obtained is called sense, and serves as the negative control.

↬ RNA polymerases are very labile enzymes that must be manipulated with extreme caution.
↬ By their nature these probes are very unstable.

1.1.5 Criteria of Choice

Table 1.1 Criteria of Choice for Determining the Type of Probe to Use

Criteria of Choice	Probes			
	Double-Stranded DNA	Single-Stranded DNA	Oligonucleotide	RNA
Availability	++	++	+++	+
Storage	+++	+++	+++	+
Stability	+++	+++	+++	+
Specific activity	++	++	++	+++
Manipulation	+++	+++	+++	+
Efficiency	+	++	++	+++
Controls	+	+	++	+++

1.2 PROBE LABELS

These labels are triphosphate nucleotides carrying radioactive isotope(s) or modified by the addition of an antigenic molecule.
The choice of label determines the type of detection method to be used:
- By autoradiography in the case of radioactive probes
- By immunohistology if the probe label has antigenic properties

The labeling can be achieved:
- After obtaining a fragment of DNA

- During the synthesis of single-stranded DNA, oligonucleotide, or RNA probes

➯ The storage conditions and duration of use are recommended by the manufacturer.
➯ Antigenic nucleotide
➯ It also dictates the sensitivity of the detection method.
➯ This method remains the most sensitive, and allows a quantitative estimation.
➯ This is a rapid method that requires neither specific equipment nor facilities.

➯ There are three methods that can be used:
- Substitution of an existing nucleotide by a labeled nucleotide
- The addition of labeled nucleotides
- The addition of molecular labels

➯ The labeled nucleotide is incorporated:
- During the synthesis, or introduced into the extension of the oligonucleotide (i.e., labeled deoxyribonucleotide triphosphates)
- During *in vitro* transcription of RNA (i.e., labeled ribonucleotide triphosphates)

1.2.1 Radioactive Labels

These are radioactive isotopes that emit β^- rays. Their position in the nucleotide structure depends on the type of isotope and the technique used for incorporation.

⇝ Radioactive labels are used exclusively in light microscopy.

1.2.1.1 Types

Of the different isotopes that can be used as labels for *in situ* hybridization, the most common ones are ^{35}S, ^{33}P, ^{32}P, ^3H, each of which has advantages and disadvantages, according to specific characteristics:
- Half-life

- Emission energy

- Resolution

- Sensitivity

- Autoradiographic efficiency

⇝ The choice of isotope depends on whether the research aim is sensitivity or resolution of detection.

⇝ Time required for half the radioactive atoms to decay.
⇝ Energy liberated when the emission of rays is evaluated in MeV (megaelectronvolts).
⇝ Distance between the localization of a silver grain and the source of emission (a function of the energy of the isotope and of the autoradiographic system).
⇝ Number of labeled molecules that can be detected (signal / background ratio).
⇝ Number of silver grains obtained by β^- radiation.

1.2.1.1.1 ^{35}S

The characteristics of ^{35}S are:
- Half-life: 87.4 days

- Emission energy: 0.167 MeV

- Resolution: 10 to 15 µm
- Sensitivity: intermediate

- Autoradiographic efficiency: 0.5 grain / β^- emission
❏ *Advantages*
- Less energetic isotope

- An excellent compromise
❏ *Disadvantages*
- Modification of the nucleotide (substitution of an oxygen molecule in the phosphate group)

⇝ ^{35}S remains the most commonly used label.

⇝ Long enough for probe use before radiolysis occurs.
⇝ Emission of β^- rays similar to ^{33}P, less energetic than ^{32}P, allowing good localization of the signal; more energetic than ^3H.
⇝ Allows cellular resolution, similar to ^{33}P.
⇝ Between ^{32}P and ^3H, similar to ^{33}P.
⇝ Specific activity 1500 Ci / mmol.
⇝ Greater than that of ^{32}P

⇝ The manipulation of ^{35}S does not require important radioprotection constraints.
⇝ Exposure times are much shorter than for ^3H (this is important in detecting weakly expressed mRNAs).
⇝ It is sensitive and efficient.

⇝ Chemical bond is less stable (*see* Figure 1.5).

Probes

- Risk of oxidation

- Difficult to use

⇝ Necessitates the use of protection (i.e., DTT, mercaptoethanol).
⇝ ^{35}S can lead to chemical modifications in the molecule, causing background.

1.2.1.1.2 ^{33}P

The ^{33}P label seems to possess the advantages of the ^{32}P label, with only minor disadvantages.
- Half-life: 25.4 days
- Emission energy: 0.25 MeV
- Resolution: 15 to 20 µm
- Sensitivity: intermediate
- Autoradiographic efficiency

❏ *Advantages*
- Weak emission energy
- Short half-life
- No modification of the nucleotide

❏ *Disadvantages*
- Cost
- Short half-life

⇝ Short exposure time
⇝ Close to that of ^{35}S
⇝ Good resolution
⇝ Close to that of ^{35}S
⇝ Close to that of ^{35}S

⇝ Few radioprotection problems
⇝ Short exposure times
⇝ Physiological label

⇝ Still high
⇝ Difficulty of storage, rapid use

1.2.1.1.3 ^{32}P

The ^{32}P label is the least often used in *in situ* hybridization
- Half-life: 14.3 days
- Emission energy: 1.71 MeV
- Resolution: 20 to 30 µm
- Sensitivity: high

- Autoradiographic efficiency: 0.25 grain / ß$^-$ emission

❏ *Advantages*
- Rapid response
- No modification of the nucleotide

❏ *Disadvantages*
- Short half-life

- High energy

⇝ Conditions of use depend upon the radioprotection constraints.
⇝ Short half-life
⇝ Very high emission energy
⇝ Poor
⇝ Detection of rare molecules in the cell
⇝ High specific activity 6000 Ci / mmol
⇝ Relatively weak

⇝ High energy, short half-life
⇝ Physiological label

⇝ Less stable (one needs to work rapidly to avoid the risks of artifacts)
⇝ Radioprotection constraints

1.2.1.1.4 ^3H

Tritiated probes lead to excellent resolution by autoradiographic detection, but the exposure time is very long (Figure 1.4).
- Half-life: 12.4 years
- Emission energy: 0.018 MeV

⇝ These probes are little used at the present time.
⇝ Radiolysis limits its use
⇝ Weak emission energy, which means that the exposure time is very long
⇝ Very weak specific activity ≈ 25 to 130 Ci /mmol.

- Resolution: 1 to 5 μm
- Sensitivity: low
- Autoradiographic efficiency: 0.20 grain / β⁻ emission

↪ Good resolution
↪ Sensitive, but the exposure time is very long
↪ Weak

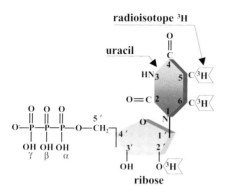

Figure 1.4 Structure of a ribonucleotide triphosphate (UTP) labeled with tritium.

❏ *Advantages*
- Possible to substitute 1 to 4 atoms of hydrogen to increase the specific activity of the probe

- No modification of the nucleotide
- Weak emission energy

❏ *Disadvantages*
- No modification of the nucleotide

- Weak emission energy / long half-life

↪ The tritium can be substituted in one or several hydrogen atoms in the base rings, or in the carbon skeleton of the sugar.
↪ This is a physiological label.
↪ There are few problems of radioprotection.

↪ There is risk of spontaneous isotope exchange.
↪ Exposure time is very long.

1.2.1.2 Position of the label

Labels (i.e., ^{35}S, ^{32}P, ^{33}P) can be placed in position α or γ according to the position of the phosphorus atoms of the nucleotide triphosphate designated α, β, γ (Figure 1.5).

↪ The choice of substituted phosphate depends on the method of labeling used:
1. Position α for
- Methods using random primer extension (*see* Section 1.3.1)
- Methods using nick translation (*see* Section 1.3.2)
- PCR (polymerase chain reaction) (*see* Section 1.3.3)
- 3' extension (*see* Section 1.3.5)
- *In vitro* transcription (*see* Section 1.3.8)
2. Position γ for
- 5' extension (*see* Section 1.3.6)

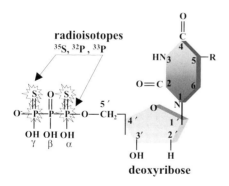

Figure 1.5 Structure of a deoxyribonucleotide triphosphate labeled by substitution of ^{35}S, ^{32}P, or ^{33}P.

- **Position α** — The phosphate placed in position α is the only one included in the phosphodiester bond. It is therefore the only one that survives in the polynucleotide chains.

⇝ Position α is used in the majority of labeling techniques.
⇝ This phosphate group creates the phosphodiester bond during polymerization (*see* Figure 1.11).

- **Position γ** — This phosphate can only be added at the 5' extremity by polynucleotide kinase.

⇝ This requires labeling by 5' extension of oligonucleotide probes (*see* Section 1.3.6).

1.2.1.3 Advantages

- Sensitivity — best results obtained with radioelements emitting less energetic, i.e., less penetrating, radiation

⇝ Radioactive probes are more sensitive than antigenic probes.
⇝ The autoradiographic signal remains located at the source of emission.

- Control of the labeling process

⇝ This allows the determination of the specific activity of the probe.

- Quantitative estimation

⇝ The density of the signal is proportional to the radiation emitted, and thus to the number of hybrids formed.

1.2.1.4 Disadvantages

- Resolution

⇝ This is poorer than that obtained with antigenic probes.

- Exposure time

⇝ After hybridization, depending upon the type of radioisotope, the slides are subjected to autoradiography.
⇝ An exposure time of between 1 day and several months is needed.

- Quantification generally necessary

⇝ Quantification is determined by the number of copies of nucleic acid in each cell. It is difficult to carry out in practice (1 mRNA molecule does not correspond to 1 grain of silver). Generally only a relative estimation is obtained.

- Use of radioisotopes requires precautions

⇝ Radioprotection constraints are needed.

- Cost of manipulations high

⇝ Renewal of the labeled nucleotides and the emulsions and labeling of the probes is required.

- Lifespan of the radioisotope limited

⇝ This requires planning of experiments.

- Radiolysis

⇝ This always remains difficult to evaluate.

1.2.2 Antigenic Labels

1.2.2.1 Types

These are nucleotide triphosphates of nitrogenous bases to which an antigenic molecule has been added by chemical coupling.
The labels most often used are biotin, digoxigenin, and fluorescein (FITC).

↪ They are used equally well in light and electron microscopy.

↪ These molecules are added by the substitution of the group – CH_3 of a thymidine. They are therefore transformed into uracil (Figure 1.7).

↪ The substitution of a methyl in position 5' allows the addition of an antigenic molecule, while transforming the dTTP into dUTP-X antigen (X corresponds to the carbon number between base and label).

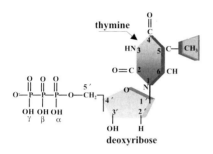

Figure 1.6 Structure of thymidine triphosphate (dTTP).

1.2.2.1.1 BIOTIN

Biotin is a vitamin (vitamin H) of low molecular weight (MW = 244).
- Affinity
  — For avidin
  — For streptavidin

- Detection
  — Avidin or multiple avidins
  — Streptavidin
  — Anti-biotin IgG (mono- or polyclonal)

↪ It is found in a number of animal tissues.

↪ Glycoprotein extract of egg white
↪ Glycoprotein extract from *Streptomyces avidinii* culture medium
↪ These molecules can be conjugated to:
  • An enzyme
  • A fluorochrome
  • Colloidal gold

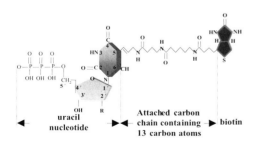

↪ Biotin – X – UTP (R = OH)
↪ Biotin – X – dUTP (R = H)
↪ X represents the number of atoms of carbon that separate the biotin from the base. It comprises between 6 and 21 carbon atoms.

Figure 1.7 Structure of a biotinylated nucleotide 16 UTP / dUTP.

❏ *Advantages*
• The biotinylated nucleotides exist in the form of:
  — Biotin – X – dUTP
  — Biotin – X – UTP

↪ The length of the chain allows the exposure of the antigenic group to the outside of the hybrid (Figure 1.7).

Probes

❏ *Disadvantages*
- Incorporation is efficient

- Biotin is present in certain tissues (e.g., muscle, liver, kidney, heart, etc.).

⇝ Very short and very long carbon chains limit hybridization.
⇝ The presence of endogenous biotin necessitates its inhibition (using blocking agents).

1.2.2.1.2 DIGOXIGENIN

Digoxigenin is a molecule extracted from the foxglove plant *(Digitalis purpurea* and *D. lanata)* of MW ≈ 300 to 400.
- Detection
  — Fab fragments
  — Anti-digoxigenin IgG (mono- or polyclonal)

⇝ This is an antigenic type molecule that does not exist in animal tissues.

⇝ A molecule which can be conjugated to:
- An enzyme
- A fluorochrome
- Colloidal gold

⇝ Digoxigenin–X–UTP (R = OH)
⇝ Digoxigenin–X–dUTP (R = H)
⇝ X represents the number of carbon atoms that separate the digoxigenin from the base. It is between 6 and 21.

1 = Structure of digoxigenin

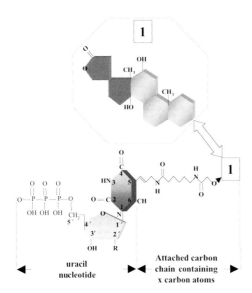

Figure 1.8 Structure of a nucleotide labeled with digoxigenin –11–UTP / dUTP.

❏ *Advantages*
- Molecule of plant origin

- Conjugated to dUTP or UTP in the form:
  — Digoxigenin – X – UTP
  — Digoxigenin – X – dUTP

- Specific detection for animals

❏ *Disadvantages*
- Structure of this molecule related to that of steroids
- Toxicity

⇝ There is no endogenous digoxigenin, as there is in animals.
⇝ X represents the carbon chain, comprising between 6 and 21 carbon atoms.
⇝ By an enzymatic reaction, the digoxigenin is attached to carbon 5 of the uracil via the attached carbon chain (Figure 1.8).

⇝ This molecule could bind to steroid receptors.
⇝ This chemical must not be inhaled.

1.2.2.1.3 FLUORESCEIN

The fluorescent label is attached directly to the nucleic acid used as the probe (Figure 1.9). Emission of a photon by fluorescein.

↪ This property of fluorescence makes it possible to visualize the label directly.
↪ This is seldom used in direct detection.

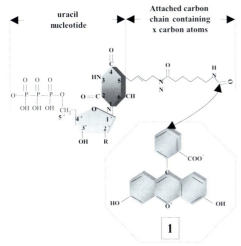

↪ Fluorescein – X – UTP (R = OH)
↪ Fluorescein – X – dUTP (R = H)
↪ X represents the number of carbon atoms that separate the fluorescein from the base. It is between 6 and 21.

1 = Structure of fluorescein

Figure 1.9 Structure of a nucleotide labeled with fluorescein.

❏ *Advantages*
• Control in probe labeling

↪ In large amounts, fluorescein allows direct visualization of the labeling quality (*see* Section 1.5.2.3).

• Antigenic probe

❏ *Disadvantages*
• Direct use very difficult for mRNA
• Sensitivity

↪ With the exception of cytogenetics
↪ Does not seem to be detectable without amplification

1.2.2.1.4 PHOTOACTIVATED MOLECULES

These are molecules of biotin or digoxigenin that can be fixed to a nucleic acid by the activation of a group under the effect of light (Figure 1.10).

↪ The storage of such an active molecule requires great care.

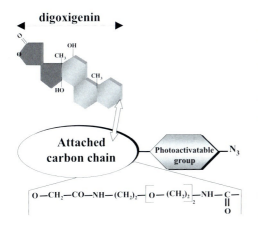

Figure 1.10 Structure of photo-digoxigenin.

Probes

❑ *Advantages*

• Fixation of the molecule onto double-stranded DNA and synthetic oligonucleotides	↬ This should not interfere with hybridization, since the reaction takes place on a double-stranded molecule in a hybrid form. At the time of hybridization, these molecules are added outside the interaction zones between the nucleic acids.
• Needs little equipment	↬ The light source must be of the right wavelength and intensity.
• Stability of labeled probes	↬ They are stable for up to 1 year.
• All the advantages of nonradioactive labels	↬ *See* Section 1.2.2.2.

❑ *Disadvantages*

• Low yield	↬ There is low specific activity.
• Cost of handling	↬ The cost is high.
• All the disadvantages of nonradioactive labels	↬ *See* Section 1.2.2.3.

1.2.2.2 Advantages

• The resolution obtained with nonradioactive probes is better than with radioactive probes.	↬ More reliable localization of the hybrids and more precise identification of subcellular structures are possible.
• The disadvantages of autoradiography are avoided.	↬ No special equipment is needed for handling.
• Results are rapid.	↬ Hybrid detection time is short.
• There are a considerable number of different labels.	↬ It is possible to see many probes carrying different labels simultaneously.
• There are numerous detection systems.	
• The probes are highly stable.	↬ They are usable for up to a year.

1.2.2.3 Disadvantages

• Sensitivity is lower.	↬ The detection of the hybrids formed at the time of hybridization necessitates the presence of a sufficient amount of antigen to permit a visible immunohistological reaction.
• Efficiency of incorporation of the labeled nucleotide varies.	↬ The efficiency of incorporation of the polymerase varies according to the length of the nucleic acid chain (a problem of steric hindrance).
• Control of probe labeling is difficult.	↬ A considerable quantity of probe and a specific detection system (*see* Chapter 5) are required.
	↬ Specific activity is not quantified with precision.
• Quantification	↬ Quantification is difficult.

1.2.3 Criteria of Choice

Table 1.2 Criteria for Determining the Label to Be Used

Criteria of Choice	Labels	
	Radioactive	**Antigenic**
Danger	Contamination	Biological
Constraints	Location / equipment	Sometimes needs to be screened from light
Cost	Periodically renewed	
Availability	Frequent	Constant
Storage of label	Short	Long
Stability	Short	Long
Efficiency of labeling	Good	Limited
Efficiency of detection	High	At the threshold of detection
Specific activity	High	Weak
Duration of the technique	Long	Rapid
Storage of the probe	Short	Long
Sensitivity	High	Limited
Resolution	At the tissue level	Cellular
Quantification	Possible	Very difficult, or even impossible

1.3 LABELING TECHNIQUES

Table 1.3 Summary of the Different Labeling Possibilities

Techniques	Probes			
	Double-Stranded DNA	Single-Stranded DNA	Oligonucleotide	RNA
Random priming	§1.3.1			
Nick translation	§1.3.2			
PCR	§1.3.3			
Asymmetric PCR		§1.3.4		
3' extension			§1.3.5	
5' extension		§1.3.6	§1.3.6	
Addition	§1.3.7	§1.3.7	§1.3.7	
In vitro transcription				§1.3.8

1.3.1 Random Priming

1.3.1.1 Principle
Random labeling by **duplication** of DNA.
Double-stranded DNA comes from the insert isolated from the plasmid.

⇝ This is accomplished by enzymatic digestion, then purification by preparative electrophoresis: migration in an agarose gel and extraction of the required fragment.

The different steps are:
- The double-stranded DNA is denatured

⇝ The insert is denatured to separate the two strands (by heat: 95°C for 5 min).

- It is incubated in the presence of a mixture of synthetic polynucleotides: **the primers.**

⇝ The polynucleotides that serve as primers are generally hexanucleotides, but can be nonanucleotides in random sequences.

- The Klenow enzyme attaches itself to the 3' hydroxylated ends of the primers and incorporates unlabeled / labeled nucleotides; these bind by complementarity to the DNA strand, which serves as the template, thus allowing the synthesis of the complementary strands.

⇝ The Klenow enzyme is the fragment of DNA polymerase I after digestion with subtilisin (disappearance of the DNase activity). This polymerase forms double-stranded DNA from a single strand and a primer.

⇝ The incorporation of the nucleotides carrying antigens is slower than with native nucleotides.

1.3 Labeling Techniques

- The α phosphate of the entering nucleotide attaches itself to the 3' hydroxyl of the deoxyribose of the preceding nucleotide to form a diester bond (Figure 1.11).

There is no degradation of the new strand.

All the newly synthesized DNA is labeled, and corresponds in total copy number to the original.

↪ The label must be carried on an α phosphate (radioactive nucleotide) (*see* Figure 1.5) or on a base (nonradioactive nucleotide) (*see* Figure 1.7).
↪ The Klenow polymerase lacks exonuclease activity.
↪ The quantity obtained seems to be larger if the reaction time is extended.

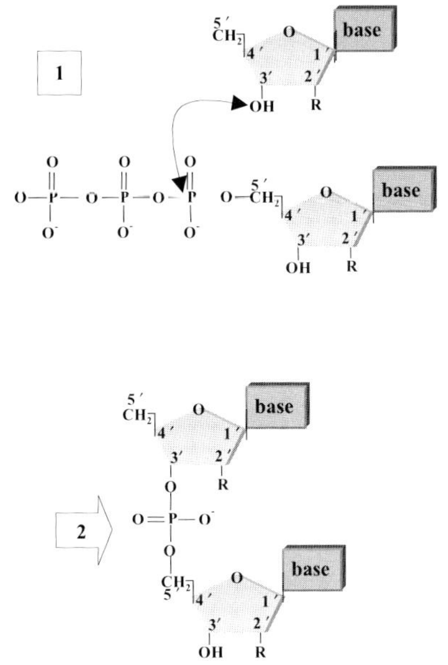

Figure 1.11 Formation of a diester bond.

❑ *Advantages*

- Specific activity — The specific activity of the labeled probe will be better if the insert is present alone.
- This labeling leads to probes of higher specific activity than that obtained by nick translation (*see* Section 1.3.2)

↪ The specific activity corresponds to the number of isotopes or antigens incorporated, in relation to the mass of the probe.
↪ The specific activity is ≈ 5 to 10 times greater than that of nick translation.
↪ The use of 1 to 4 labeled nucleotides adds proportionally to the specific activity of the probe, but this generally increases radiolysis in the case of radioactive nucleotides. Generally, a single labeled nucleotide is used, or sometimes two.
↪ The ratio of labeled (antigenic) to unlabeled nucleotides has to be high to facilitate their incorporation (a ratio of 1/4 is generally advised).

- There is uniform labeling of DNA.
- Only a small amount of DNA is necessary to achieve labeling.

↪ Labeling can be achieved with a quantity of DNA ≥ 10 ng.

Probes

❏ *Disadvantages*

• Preparation of the DNA is longer than with the nick translation method.
• Smaller generated DNA fragments than with nick translation.

↪ It is necessary to use the insert rather than the entire plasmid.
↪ The generated fragments are of the order of 100 to 200 nucleotides in length.

1.3.1.2 Summary of the different steps

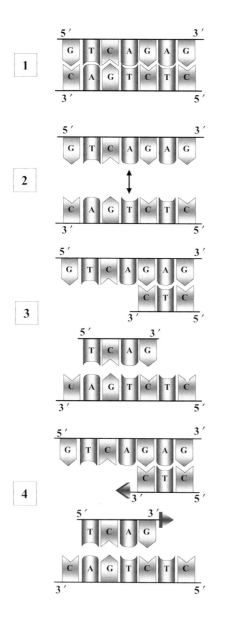

1 = Separation of the insert from the plasmid — The labeling is achieved after separation of the insert from the plasmid (vector) by enzymatic cutting and agarose gels.

2 = Denaturation — The DNA is double-stranded; **denaturation** (separation of the double strands) is an **indispensable stage** in the hybridization of primers.

3 = Hybridization of the primers — Hybridization with oligonucleotides whose sequences contain all the possible combinations: the type of primer (hexa- or nonanucleotide) dictates the length of the newly synthesized strands. The label must be carried by the α phosphate or the base.

4 = Binding of the Klenow enzyme ➤ Binds to the 3' hydroxylated ends, which then incorporate the labeled or nonlabeled precursors.

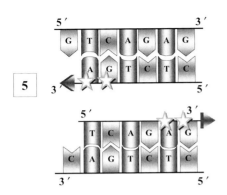

5 = Incorporation of nucleotides labeled with dGTP or dATP, and unlabeled nucleotides
- Synthesis is obtained by the action of the Klenow enzyme, which incorporates the nucleotides by copying the complementary strand.
- The length of the primers influences the length of the newly formed strands.
- Polymerization is effective in the **5' → 3'** sense.

6 = Denaturation
- Before use, the fragments must be denatured.
- The labeled fragments are from 100 to 200 nucleotides in length.

Figure 1.12 Labeling of DNA by random priming.

1.3.1.3 Equipment/reagents/solutions

1. Equipment
- Water bath or heating block **100°C** / **37°C**
  - ↪ Denaturation of DNA
  - ↪ Incubation of the enzyme
- Centrifuge
  - ↪ ≥ 14,000 g
- –20 or –80°C freezer
  - ↪ Storage of products and precipitation of the probe
- Vortex mixer
  - ↪ Should be used carefully so as not to disperse the products on the wall of the tube

2. Reagents
- Dithiothreitol (DTT)
  - ↪ Only used for *in situ* hybridization
- EDTA
  - ↪ Ethylene diamine tetra-acetic acid
- Hexanucleotides
  - ↪ Or nonanucleotides

3. Solutions
  - ↪ **All the solutions are prepared using RNase-free reagents in a sterile container.** (*see* Appendix A2).
- DNA to be labeled
  - ↪ The DNA is between 400 bp and several kbp.
  - ↪ The typical final concentration for labeling is 1 µg / µL (100 ng to 3 µg).
  - ↪ Store at –20°C in glycerol 1:1 (v/v).
- Klenow enzyme 2 U / µL
  - ↪ Concentration should be 10 to 50 ng.
- Probe
  - ↪ This is generally supplied with the enzyme.
- Buffer 10X or reaction buffer containing hexanucleotides 10X
  - ↪ This is provided with the labeling kit.
- dNTP (dATP, dTTP, dCTP, dGTP) 10X in 500 mM Tris–HCl buffer containing 50 mM MgCl$_2$, 1 mM DTT, BSA 0.5 mg/mL; pH 7.8
  - ↪ Store at –20°C.
  - ↪ Use DNase-free BSA.

- Labels
  — Antigenic labeling

  — Antigenic labels:
  1 mM biotin – 11 – dUTP
  1 mM digoxigenin – 11 – dUTP
  1 mM fluorescein – 12 – dUTP
  — Radioactive labeling

  — Radioactive labels:
  $\alpha\,^{35}$S / dCTP, $\alpha\,^{33}$P / dCTP
  $\alpha\,^{35}$S / dATP, $\alpha\,^{33}$P / dATP
- Sterile water
- Dithiothreitol 1 M
- 200 mM EDTA; pH 8.0

↪ Concentration is 1 mM (dATP, dCTP, dGTP), 0.65 mM (dTTP) + 0.25 mM labeled dUTP.
↪ Store in aliquots at –20°C.

↪ Concentration of radioactive or nonradioactive is 0.5 mM dNTPs (dATP, dTTP, dCTP, dGTP).
↪ Store in aliquots at -80°C.

↪ See Appendix B1.1.
↪ See Appendix B2.10.
↪ See Appendix B2.12.

1.3.1.4 Protocol for radioactive probes

1. *Denaturation*

a. Place the following reagents in a sterile Eppendorf tube in the given order:
- Distilled sterile water   9 µL
- Probe   2 µL

b. Denature.   **10 min at 100°C**

c. Place immediately in   **10 min at –10°C**
- Ice / NaCl
or
- Ice / ethanol

↪ The labeling uses single-stranded DNA.

↪ The typical final concentration for labeling is 0.5 µg / µL.

↪ Temperatures above 92°C for several minutes allow the separation of the two strands of DNA.

↪ To prevent reassociation at the time of cooling, the solution is placed very rapidly into a bath below 0°C (the two strands remain separated).

2. *Duplication by the Klenow enzyme*

a. Add:
- 0.5 mM dNTP   3 µL
  (dCTP, dGTP, dTTP)
- Probe   1 µL
- Labeled dATP (50 µCi)   5 µL
- Hexanucleotides 10X   2 µL
- Klenow enzyme   **1 µL (+ 1 µL)**

↪ 1 µL of each unlabeled nucleotide.

↪ 25 ng of denatured, double-stranded DNA.
↪ Another labeled dNTP can be used.
↪ Hexa- or nonanucleotide primers.
↪ Final concentration: 2 units. This concentration must be modified to balance the overall quantity of probe to be labeled.
↪ Because of the instability of the newly synthesized strands, it is possible to add more of the enzyme to increase the quantity of DNA produced. It should be added at the last moment.

• Sterile water	to 20 μL	↪ See Appendix B1.
b. Incubate.	60 min at 37°C	↪ The reaction can be continued for up to 20 h.
		↪ Note: enzyme is destroyed after 90 min of incubation. Add more if necessary.
c. Add EDTA 200 mM.	2 μL	↪ Stops the reaction
		↪ The same effect can be obtained by heating for 10 min at 65°C.

3. *Separation of the nucleotides* not incorporated by chromatography

↪ Pass through a Sephadex G50 column or Nuctrap push column (sometimes not necessary); *see* Section 1.4.2.

Next step — Purification of the DNA by ethanol precipitation.

↪ *See* Section 1.4.1.

1.3.1.5 Protocol for antigenic probes

1. *Denaturation*
↪ *See* Section 1.3.1.4.

2. *Duplication by the Klenow enzyme*
  a. Mix the unlabeled nucleotides

↪ Prepare a solution of unlabeled nucleotides by the addition of an equal volume of each nucleotide.

• dNTP mixture	2 μL	↪ Final concentration (25 nM dATP, dCTP, dGTP, 16 nM dTTP).
• Labeling buffer	2 μL	↪ Initial concentration 10X.
• Labeled dNTPs	1 μL	↪ (Biotin, digoxigenin, fluorescein) – dUTP used at a final concentration of 9 nM.
• Hexanucleotides 10X	2 μL	↪ Hexa- or nonanucleotide primers.
• Klenow enzyme	1 μL (+ 1 μL)	↪ Final concentration: 2 units.
		↪ Because of the instability of the newly synthesized strands, it is possible to add more enzyme to increase the quantity of DNA formed. This should be added at the last moment.
• Sterile water	to 20 μL	↪ *See* Appendix B1.
b. Incubate.	60 to 90 min at 37°C	↪ The reaction can be continued for up to 20h.
		↪ Note: The enzyme is destroyed after 90 min of incubation. Add more if necessary.
		↪ In the polymerase reaction, one nucleotide in every 20 to 25 is labeled. The incubation time is therefore longer (given the problem of steric hindrance).
c. Add EDTA 200 mM.	2 μL	↪ The same effect can be obtained by heating for 10 min at 65°C.

Next step — Purification of the DNA by ethanol precipitation.

↪ *See* Section 1.4.1.
↪ Do not use the method of separation on a Sephadex column because of the problem of steric hindrance.

1.3.2 Nick Translation

1.3.2.1 Principle

Double-stranded DNA is incubated in the presence of DNase I. This enzyme generates random nicks in each of the DNA strands.

↪ The enzyme cuts the phosphodiester bonds randomly. The number of breaks depends on the concentration of DNase. Use the concentration given in the kits.

The DNA polymerase attaches itself at the point of the breaks. Its exonuclease activity eliminates the nucleotides from the newly created 5' end, and its polymerase activity adds new nucleotides to the 3' OH end.

↪ The two properties of the DNA polymerase I used are (1) 5' → 3' exonuclease activity (degradation activity: the enzyme eliminates the exposed 5' nucleotides); (2) 5' → 3' polymerase activity (activity of polymerization 5' → 3' of the DNA polymerase I, which replaces the missing nucleotides).

The DNA polymerase does not have ligase activity; a succession of short fragments is therefore synthesized.

↪ There is no reconstitution of the entire molecule.
↪ One or several nucleotides are labeled (radioactive or antigenic labels); the newly synthesized strands are labeled.

The nucleotides carry:
• Radioactive isotope(s)

↪ ^{33}P, ^{35}S, ^{3}H, ^{32}P
• If labeling with ^{32}P (>3000 Ci / mmol), use just one radioactive nucleotide.
• For ^{3}H labeling (>25 to 30 Ci / mmol), use several radioactive deoxyribonucleotides.

• An antigenic molecule

↪ For example:
• Biotin (dUTP – 11 – biotin)
• Digoxigenin (dUTP – 11– digoxigenin)
• Fluorescein (dUTP – 12 – fluorescein)

❏ *Advantages*
• Rapid method

↪ No preliminary treatment of the stored DNA is necessary. The DNA in the plasmid can be used directly, without separating the insert.

• Biological stability

↪ The enzymes of degradation of the DNA are very unstable, and are effectively inhibited by chelating agents such as EDTA.

❏ *Disadvantages*
• DNA polymerase lacking ligase activity on the ends
• Specific activity

• Quantity of DNA necessary to achieve labeling

↪ The probe corresponds to the fragments of labeled DNA.
↪ Specific activity is weaker than with random priming (*see* Section 1.3.1)
↪ The labeling can be achieved with a quantity of DNA > 150 to 200 ng (10 times greater than random priming (*see* Section 1.3.1).

1.3 Labeling Techniques

- Use

⇝ This method is little used, as it has few advantages.

1.3.2.2 Summary of the different steps

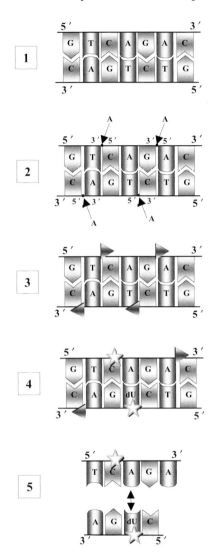

1 = Separation of the insert from the plasmid.

2 = Enzymatic digestion — Incubation in the presence of DNase, which generates random nicks (A) in each of the strands of DNA.

3 = Enzyme action ➤ Two properties of the DNA polymerase are used:
- Exonuclease activity 5' → 3'
- Polymerase activity 5' → 3'

4 = Incorporation of the nucleotides labeled with dCTP and dUTP:
- In the course of the polymerase action, the labeled nucleotides (*) are incorporated.
- The exonuclease activity eliminates the nucleotides downstream from the break.

5 = Synthesis of short fragments:
- The absence of ligase activity prevents the reconstitution of the entire molecule.
- The fragments of the probe are denatured before use.

Figure 1.13 Labeling of DNA by nick translation.

1.3.2.3 Equipment/reagents/solutions
1. Equipment
- Refrigerated water bath (**15°C**)

⇝ Optimal temperature for the enzyme (A temperature of 15°C is obtained in a Dewar by placing it in running water; adjust to 10°C with ice. Check the temperature after 90 min, and readjust if necessary.)

- Centrifuge
- Vacuum jar or Speedvac

⇝ ≥14,000 g
⇝ To dry the probe

Probes

- −20 and / or −80°C freezer ↪ For storage of the products, and precipitation of the probe
- Vortex mixer ↪ Should be used carefully so as not to disperse the products on the wall of the tube

2. Reagents
↪ Molecular biology quality, only to be used for *in situ* hybridization

- Magnesium chloride ($MgCl_2$)
- Enzymes: ↪ Enzymes contained in the labeling kit
  — DNA polymerase I
  — DNase I
- Nucleotides: ↪ Generally 100 mM
  — Unlabeled nucleotides ↪ Reagents in the labeling kit
  — Antigenic nucleotides X – dUTP ↪ Biotin, digoxigenin, or fluorescein
- Labeling buffer 10X ↪ Buffer in the labeling kit

3. Solutions
↪ **All the solutions are prepared with RNase-free reagents in a sterile container** (*see* Appendix A2).

- Sterile water ↪ *See* Appendix B1.1.
- Enzymes: ↪ Store at −20°C.
  — DNA polymerase I   10 U / µL
  — DNase I   0.1 ng / µL ↪ *See* Appendix B2.11.
- 50 mM $MgCl_2$ ↪ *See* Appendix B2.7.
- Nucleotides:
  — Unlabeled nucleotides ↪ 0.5 mM dNTP, store at −20°C.

  — Labeled nucleotides:
    Antigenic nucleotides – X – dUTP ↪ X: *see* Figure 1.7.
  ↪ Concentration determined by the labeling kit or independently (1 nM)

    Radioactive nucleotides ↪ Generally 5 µCi
- Nick translation buffer 10X ↪ Labeling buffer; use that provided in the kit
  — 500 mM Tris – HCl; pH 7.8
  — 50 mM $MgCl_2$ ↪ *See* Appendix B3.6.
  — 100 mM DTT ↪ *See* Appendix B2.7.
  — BSA 0.5 mg / mL ↪ *See* Appendix B2.10.

1.3.2.4 Protocol for radioactive probes

↪ DNA probe labeled with:
- $\alpha(^{35}S, ^{33}P, \text{ or } ^{32}P)$ – dNTP
- Mixture of ^3H – dNTP

1. *Lyophilization of the radioactive nucleotides* ↪ Optional
  a. Place in a sterile tube ↪ Eppendorf-type
  • Labeled dNTP   X µL ↪ Labeled nucleotides ≈ 1000 Ci / mmol
  b. Evaporate on a Speedvac   45 min ↪ Recommended

2. *Reaction mixture*
  a. Add to the dry residue
  • Sterile water   36 µL
  • Nick translation   5 µL
    buffer 10X

1.3 Labeling Techniques

- 2 nM unlabeled dNTP / µL   **5 µL/nucleotide**
- Labeled dNTP   **X µL**
- Plasmid DNA ≈ 0.1 to 3 µg / µL   **1 µL**

b. Vortex, centrifuge.   ≈ **1000 rpm at 4°C**

⇝ If $\alpha^{35}S$ – dCTP is the label, use the mixture of dGTP, dTTP, dATP.
⇝ Use 1 to 5 µL of each labeled nucleotide.
⇝ To recover all the constituents at the bottom of the tube

3. Enzymatic digestion —
DNase I   **1 µL**
a. Mix.
b. Incubate.   **4 min at 37°C**

⇝ 5 ng / µL
⇝ Gently
⇝ The time varies depending on the batch of DNase I used.

4. Enzymatic action
a. Place on ice immediately and add DNA polymerase I 10 U / µL.   **1 µL**

b. Incubate.   **90 min to 3 h at 15°C**

c. Add 10 mM EDTA.   **2 µL**

d. Vortex, centrifuge.

⇝ Mix gently; do not vortex.

⇝ Incorporation of the labeled nucleotides: after 2 h at 15°C all the missing bases are replaced.
⇝ This stops the reaction.
⇝ The same effect can be obtained by heating for 10 min at 65°C.
⇝ Mix, withdraw X µL to measure the total radioactivity before extraction of the probe.
⇝ Stopping the reaction is **not essential**. This can be replaced by passage through a column.

Next step — Purification of the DNA by ethanol precipitation.

⇝ *See* Section 1.4.1.

1.3.2.5 Protocol for antigenic probes

⇝ DNA probe labeled with (biotin, digoxigenin, fluorescein) – dUTP

1. Reaction mixture
a. Add to the dry pellet the following reagents:
- Sterile water   **20 µL**
- Nick translation buffer 5X   **4 µL**

- Unlabeled dNTPs   **1 µL/nucleotide**

- Labeled dUTP   **2 µL**
- Plasmid DNA ≈ 0.1 to 3 µg / µL   **X µL**

b. Vortex, centrifuge.   ≈ **1000 rpm at 4°C**

⇝ In the order indicated
⇝ Certain mixtures contain the buffer, the unlabeled nucleotides, and the enzymes
⇝ Concentration (0.5 mM dATP, dCTP, dGTP, and 0.1 mM dTTP)
⇝ Concentration: 1 mM labeled dUTP
⇝ ≈ 1 µg probe
⇝ To recover all the constituents at the bottom of the tube

2. Enzymatic digestion — DNase I   **1 µL**
a. Mix.
b. Incubate.   **4 min at 37°C**

3. Enzymatic action
a. Place immediately on ice and add DNA polymerase I 10 U / µL.   **2 µL**

⇝ 5 ng / µL
⇝ Gently
⇝ To control the frequency of strand breaks

⇝ Mix gently

b. Incubate. **90 min to 3 h at 15°C**
  ↪ Incorporation of the labeled nucleotides: after 2 h all the missing bases are replaced.

c. Add 10 mM EDTA. **X µL**
  ↪ This stops the reaction.
  ↪ It is possible to stop the reaction by heating for 10 min at 65°C.

d. Vortex.
  ↪ To mix

Next step — Purification of the DNA by ethanol precipitation.
  ↪ *See* Section 1.4.1.

1.3.3 Amplification by PCR

↪ Polymerase chain reaction

1.3.3.1 Principle

To amplify a specific sequence of a given nucleic acid (e.g., a probe) in a mass of DNA.

↪ The quantity obtained is proportional to the number of cycles used (2^n, where n is the number of cycles).

The probe and its complementary strand are present in the midst of a mass of nucleic acid (e.g., plasmid DNA, *see* Figure 1.3). Two primers define this sequence: the first is "sense" at the 5' extremity to the probe, the second "anti-sense" at the 3' extremity to the probe sequence (Figure 1.14). After hybridization of these oligonucleotides, duplication is achieved by the action of a polymerase, which extends the oligonucleotides by using the cDNA as a template.

A large copy number is obtained by a series of cycles comprising a primer hybridization stage, followed by a stage of elongation of this primer, and finally a denaturation stage which separates the newly formed strand from the original template strand.

↪ The *Taq* enzyme withstands the temperature of denaturation.

❑ *Advantages*
- Production of a large amount of probe
- No pretreatment of the stored DNA
- Method of amplification of a specific sequence
- Labeling during the course of synthesis
- Newly synthesized probe able to reach several hundred bases
- High specific activity
- High specificity of the probe
- DNA probe

↪ Great stability of DNA probes

❑ *Disadvantages*
- Requires special equipment
- Very stringent reaction conditions
- Considerable risk of contamination

↪ Loss of probe specificity

1.3.3.2 Summary of the different steps

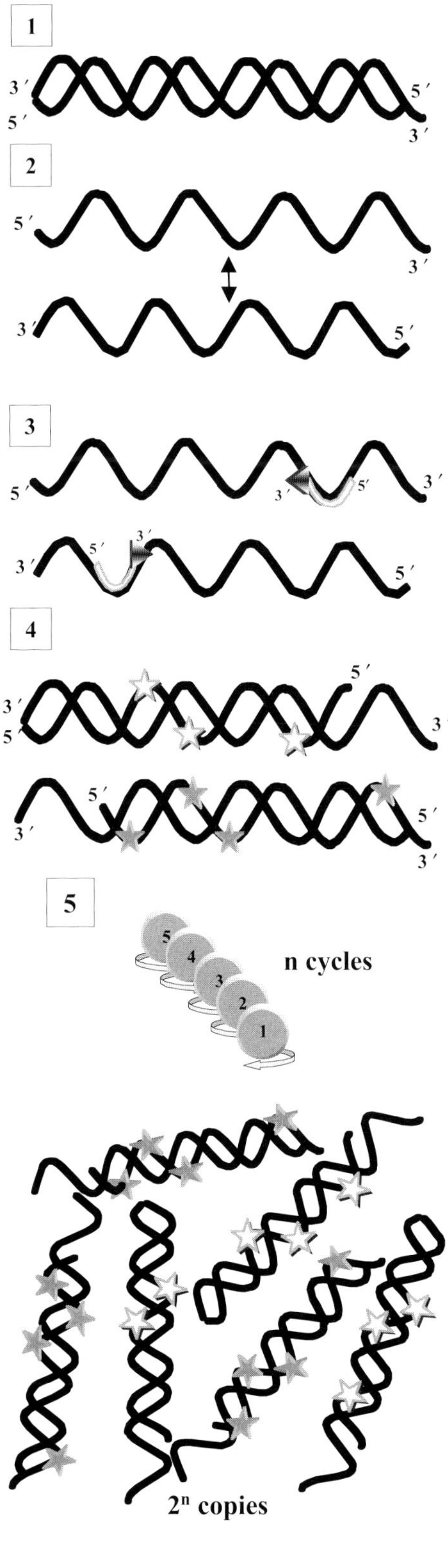

1 = **Genomic DNA**.

2 = **Denaturation** — Genomic DNA is denatured.

3 = **Addition of primers** — The primer specific for each end of the two strands of the sequence to be amplified is hybridized.

4 = **Action of *Taq* polymerase** — The enzyme extends the primer by duplication of the DNA strand (end of the first cycle).

n cycles

5 = **Amplification** — The following cycles lead to the amplification of the target sequence, and therefore the probe. The labeled nucleotides are incorporated at the time of synthesis of the newly formed strand. The quantity of the probe obtained is proportional to the number of cycles.

2^n copies

Figure 1.14 Labeling by symmetrical PCR.

1.3.3.3 Equipment/reagents/solutions

1. Equipment — Liquid PCR machine ↪ Standard equipment

2. Reagents
- Sense and anti-sense primers ↪ Storage in aliquots at –20°C
- Labeled dNTP:
  — dUTP – X – antigen ↪ Storage at –20°C
  — Radioactive dNTP ↪ Storage at 4°C or –80°C
- Unlabeled dNTPs ↪ Storage at –20°C
- *Taq* polymerase enzyme ↪ Storage at –20°C
- Mineral oil ↪ For PCR
- KCl ↪ Molecular biology grade
- $MgCl_2$ ↪ Molecular biology grade
- Tris ↪ Molecular biology grade

3. Solutions ↪ **All the solutions are prepared using DNase-free reagents in a sterile container** (*see* Appendix A2).

- Sense and anti-sense primers ↪ 0.1 to 1 μM (stored in aliquots at –20°C)
- Labeled dNTP ↪ Storage at –20°C
  — 0.7 mM dUTP – X – antigen ↪ Addition of 1.3 mM dTTP
  — Radioactive dNTP
- 2 mM unlabeled dNTPs ↪ Storage at –20°C
- Sterile water ↪ *See* Appendix B1.
- *Taq* polymerase enzyme ↪ 5 U / μL (storage at –20°C)
- 25 mM $MgCl_2$ ↪ *See* Appendix B2.7.
- Buffer 10X ↪ 100 mM Tris – HCl; 500 mM KCl; pH 8.3
  ↪ Storage in aliquots at –20°C

1.3.3.4 Reaction mixture for radioactive labeling

1. Place the following reagents in a sterile Eppendorf tube:

• Linearized DNA	≈ 50 ng	↪ To be determined
• Primers	250 nmol	↪ Of each primer
• 3 unlabeled dNTPs	X μL	↪ 2 mM
• 4th labeled dNTP	X μL	↪ ≈ 50 μCi
• 4th unlabeled dNTP	X μL	↪ ≈ 1 mM
• $MgCl_2$	2 to 10 μL	↪ To be determined
• Buffer 10X	5 μL	
• *Taq* polymerase	1.5 U	↪ To be determined (0.5 to 2.5 U)
		↪ Volume according to the concentration
• H_2O	to 50 μL	↪ *See* Appendix B1.

2. Mix and centrifuge.
3. Cover with oil. 100 μL
4. Place in the thermocycler.

1.3.3.5 Reaction mixture for antigenic labeling

1. Place the following reagents in a sterile Eppendorf tube:

• Linearized DNA	≈ 50 ng	↝ To be determined
• Primers	250 nmol	
• dATP, dGTP, dCTP	X µL	↝ 2 mM
• Labeled dUTP	X µL	↝ 0.7 mM
• Unlabeled dTTP	X µL	↝ 1.3 mM
• MgCl$_2$	2 to 10 µL	↝ To be determined
• Buffer 10X	5 µL	
• *Taq* polymerase	1.5 U	↝ To be determined (0.5 to 2.5 U)
		↝ Volume according to the concentration
• H$_2$O	to 50 µL	

2. Mix and centrifuge.

3. Cover with oil.   100 µL

4. Place in the thermocycler.

1.3.3.6 PCR protocol

1. *First cycle*

• Denaturation	7 min at 94°C	
• Hybridization	1 min at 60°C	↝ This temperature varies according to the primer.
• Elongation	1 min at 72°C	↝ The time can be extended if the probe to be synthesized is long.

2. *Following cycles* (*n* cycles)

↝ The number depends on the required quantity of probe: $Q = q \times 2^n$, where q is the quantity of probe at the start, and n the number of cycles.

• Denaturation	1 min at 94°C	↝ For large numbers of cycles, it is sometimes necessary to reduce this time to preserve the efficiency of the enzyme.
• Hybridization	1 min at 60°C	↝ This temperature varies according to the primer.
• Elongation	1 min at 72°C	↝ After 10 to 20 cycles, this time is generally increased to compensate for the loss in efficiency of the enzyme.

3. Last cycle

• Denaturation	1 min at 94°C	
• Hybridization	1 min at 60°C	
• Elongation	10 min at 72°C	↝ Time is increased to complete the extension of the newly formed strands.

Next step — Precipitation with ethanol.   ↝ *See* Section 1.4.1.

1.3.4 Amplification by Asymmetric PCR

The aim is to compensate for the principal drawback of the classical PCR by obtaining the probe in the form of two complementary strands, i.e., by generating the anti-sense sequence of the target nucleic acid.

1.3.4.1 Principle

The use of a single primer ensures the synthesis of the probe to be used. After denaturation of the two strands of DNA (first step), the primer hybridizes to the target DNA strand (second step) then, during the course of the third step, elongation (i.e., probe synthesis) takes place. In each cycle, an anti-sense copy of the target nucleic acid is synthesized. The number of copies obtained is n (where n represents the number of cycles).

↝ This proportionality is no longer exponential, but arithmetical.

❏ *Advantages*
- Production of the probe alone
- Single-stranded DNA probe
- Labeling during the course of synthesis
- High specific activity
- No pretreatment of the stored DNA
- Newly synthesized probe can reach several hundred bases

↝ A feature of the method
↝ High stability of DNA probes

❏ *Disadvantages*
- Weak yield
- Necessitates specific equipment
- Very stringent reaction conditions
- Considerable risk of contamination

↝ Quantity of probe obtained is limited.

↝ Loss of probe specificity

1.3.4.2 Summary of the different steps

1

1 = Genomic DNA.

1.3 Labeling Techniques

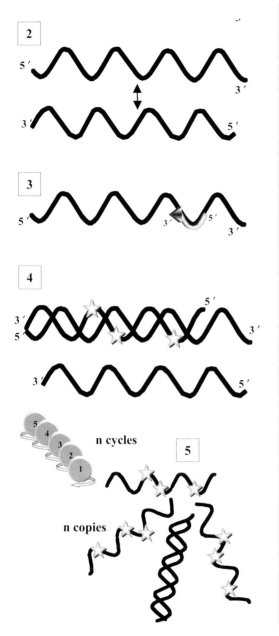

2 = Denaturation — Genomic DNA is denatured.

3 = Addition of primer — The primer is specific to the end of the strand that is complementary to the sequence to be detected.

4 = Action of *Taq* polymerase — The enzyme extends the primer by duplication of the DNA strand (end of the first cycle).

5 = Amplification — The following cycles lead to the amplification of the probe. The labeled nucleotides are incorporated at the time of synthesis of the newly formed strand. The quantity of probe obtained is proportional to the number of cycles.

Figure 1.15 Labeling by asymmetric PCR.

1.3.4.3 Equipment/Reagents/Solutions

1. *Equipment* — Liquid PCR machine

↪ Standard for PCR

2. *Reagents*

- Anti-sense primer

↪ The only primer that allows the synthesis of the sequence that is complementary to the target
↪ Storage in aliquots at −20°C

Probes

- Labeled dNTP
- dUTP – X – antigen
- Radioactive dNTP
- Unlabeled dNTPs
- *Taq* polymerase enzyme
- Mineral oil
- KCl
- MgCl$_2$
- Tris

⇝ Storage at –20°C
⇝ Storage at 4°C or –80°C
⇝ Storage at –20°C
⇝ Storage at –20°C
⇝ For PCR (stored at –20°C)
⇝ Molecular biology grade
⇝ Molecular biology grade
⇝ Molecular biology grade

3. *Solutions*

⇝ **All the solutions are prepared using DNase-free reagents in a sterile container** (*see* Appendix A2)

- Anti-sense primer
- Labeled dNTP
  — 0.7 mM dUTP – X – antigen
  — Radioactive dNTP
- 2 mM unlabeled dNTPs
- Sterile water
- *Taq* polymerase enzyme
- 25 mM MgCl$_2$
- Buffer 10X

⇝ 0.1 to 1 µM (storage in aliquots at -20°C)
⇝ Stored at –20°C
⇝ Addition of 1.3 mM dTTP

⇝ Stored at –20°C
⇝ *See* Appendix B1.1.
⇝ 5 U / µL (stored at –20°C)
⇝ *See* Appendix B2.7.
⇝ 100 mM Tris – HCl; 500 mM KCl; pH 8.3
⇝ Storage in aliquots at –20°C

1.3.4.4 Reaction mixture for radioactive labeling

1. Place the following reagents in a sterile Eppendorf tube:

• Linearized DNA	≈ **50 ng**
• Primer	**250 nmol**
• 3 unlabeled dNTPs	**X µL**
• 4th labeled dNTP	**X µL**
• 4th unlabeled dNTP	**X µL**
• MgCl$_2$	**2 to 10 µL**
• Buffer 10X	**5 µL**
• *Taq* polymerase	**1.5 U**
• H$_2$O	**to 50 µL**

⇝ To be determined
⇝ Only the anti-sense primer
⇝ A very weak concentration of the sense primer (< 50 to 100 times)
⇝ 2 mM
⇝ ≈ 50 µCi
⇝ ≈ 1 mM
⇝ To be determined

⇝ To be determined (0.5 to 2.5 U)
⇝ Volume according to the concentration
⇝ *See* Appendix B1.

2. Mix and centrifuge.

3. Cover with oil. **100 µL**

4. Place in the thermocycler.

1.3.4.5 Reaction mixture for antigenic labeling

1. Place the following reagents in a sterile Eppendorf tube:

• Linearized DNA	≈ 50 ng	↪ To be determined
• Primer	250 nmol	↪ Only the anti-sense primer
• dATP, dGTP, dCTP	X µL	↪ 2 mM
• Labeled dUTP	X µL	↪ 0.7 mM
• Unlabeled dTTP	X µL	↪ 1.3 mM
• MgCl$_2$	2 to 10 µL	↪ To be determined
• Buffer 10X	5 µL	
• *Taq* polymerase	1.5 U	↪ To be determined (0.5 to 2.5 U)
		↪ Volume according to the concentration
• H$_2$O	to 50 µL	↪ *See* Appendix B1

2. Mix and centrifuge.
3. Cover with oil.   **100 µL**
4. Place in the thermocycler.

1.3.4.6 Protocol for asymmetric PCR

1. ***First cycle***

• Denaturation	7 min at 94°C	↪ The time can be reduced.
• Hybridization	1 min at 60°C	↪ This temperature varies according to the primer.
• Elongation	1 min at 72°C	↪ The time can be increased if the probe to be synthesized is long.

2. ***Following cycles***

↪ The number depends on the required quantity of probe: Q = q × n, where q is the quantity of probe at the start and n the number of cycles.

• Denaturation	2 min at 94°C	↪ For large numbers of cycles it is sometimes necessary to reduce this time to preserve the efficiency of the enzyme.
• Hybridization	1 min at 60°C	↪ This temperature varies according to the primer.
• Elongation	1 min at 72°C	↪ After 10 to 20 cycles, this time is generally increased to compensate for the loss in the efficiency of the enzyme.

3. ***Last cycle***   ↪ **This is a particularly important cycle.**

• Denaturation	2 min at 94°C	
• Hybridization	1 min at 60°C	
• Elongation	10 min at 72°C	↪ Time is increased to complete the extension of the newly formed strands.

Next step — Precipitation with ethanol.   ↪ *See* Section 1.4.1.

1.3.5 3' Extension

1.3.5.1 Principle
The labeling of an oligonucleotide (<50 mers) is carried out by the addition at the 3' end of labeled nucleotides using an enzyme, the terminal transferase. This requires a free 3' OH and nucleotide triphosphates.

↪ Method is applicable to oligonucleotides.

↪ The methods of random primer extension (*see* Section 1.3.1), nick translation (*see* Section 1.3.2), and PCR (*see* Sections 1.3.3 and 1.3.4) are not applicable to these probes.
↪ It is possible to add unlabeled nucleotides, so as to have a longer extension in the case of antigenic nucleotides.

In the presence of cobalt ions, the enzyme (terminal deoxytransferase, TdT) catalyzes the polymerization of a labeled deoxyribonucleotide.
The extension poly (T) is to be avoided.

↪ Cobalt is the cofactor of TdT.

↪ A poly (A) end is present in the messenger RNA of eukaryotic cells, which would form hybrids.

The labeled nucleotides are carriers of:
- Radioactive isotopes: ^{35}S or ^{33}P
- An antigenic molecule

↪ α (^{35}S or ^{33}P) – dATP
↪ Such as:
- Biotin – X – dUTP
- Digoxigenin – X – dUTP
- Fluorescein – X – dUTP

❏ *Advantages*
- Use in the case of strong homologies

↪ The synthesis of oligonucleotides for specific regions

- Can be used to label the breaks in genomic DNA
- Single-stranded probe
- Little DNA required
- Small fragments
- Sensitivity

↪ No reassociation possible with the probe
↪ ≥10 pmol
↪ Easy penetration in the tissue
↪ Possible to improve the sensitivity of detection to the level of detecting mRNA by simultaneously using several labeled oligonucleotides to detect the same target nucleic acid.

- Rapid method
- Radioactive labeling

↪ No manipulation by cloning
↪ Determined by specific activity, which can be very high

- Nonradioactive labeling

↪ Possible to obtain large quantities of DNA for storage

- Quantification

↪ Each hybrid corresponding to a target nucleic acid

❏ *Disadvantage*
- Weak signal using antigenic labeling

↪ Each oligonucleotide corresponds to a target nucleic acid. If the expression is weak, the signal is weak. It can be increased by using several probes for the same target sequence.

1.3.5.2 Summary of the different steps

1 = Synthesized oligonucleotide probe—The 3' end must be hydroxylated.

2 = Enzymatic action ➤ Terminal deoxytransferase (TdT) attaches itself to the 3' end containing a free OH.

3 = Addition of labeled dNTPs (dATP) and unlabeled dNTPs, and polymerization — The TdT enzyme catalyzes the polymerization of the radioactive and antigenic deoxynucleotide triphosphates.

Figure 1.16 Labeling by 3' extension.

1.3.5.3 Equipment/reagents/solutions

1. Equipment
- Water bath / incubator at 37°C
- Centrifuge
- Vortex mixer

↪ Enzyme reaction
↪ ≥14,000 g

2. Reagents

- Potassium cacodylate
- Cobalt chloride ($CoCl_2$)
- EDTA
- Antigenic dUTP labels:
  — Biotin – X – dUTP
  — Digoxigenin – X – dUTP
  — Fluorescein – X – dUTP
- Radioactive dNTP labels

- Oligonucleotides

- Terminal deoxytransferase (TdT)

↪ Molecular biology grade
↪ Only to be used for *in situ* hybridization

↪ Labeling kit

↪ Essentially $\alpha(^{35}S$ or $^{33}P) - dATP$, the other radioisotopes being little used (the use of other dNTPs leads to nonspecific signals)
↪ 25 to 45 nucleotides (mers) (a probe of 30 nucleotides is a good average)
↪ Storage at –20°C
↪ The enzyme sold with its buffer and $CoCl_2$

3. Solutions

- $CoCl_2$ 25 mM
- Antigenic deoxynucleotides

↪ **All the solutions are prepared using DNase- and RNase-free reagents in a sterile container** (*see* Appendix A2).
↪ Storage at –20°C (*see* Appendix B2.5.2)
↪ Storage at –20°C

- 1 mM biotin – X – dUTP
- 1 mM digoxigenin – X – dUTP
- 1 mM fluorescein – X – dUTP
- Radioactive deoxynucleotides
- $\alpha(^{35}S$ or $^{33}P) - dATP$

- Water
- 200 mM EDTA; pH 8.0
- Oligonucleotides 2 to 100 pmol / µL
- Labeling (tailing) buffer 5X

- Terminal deoxytransferase (TdT) 50 U / µL

↪ X: *see* Figure 1.7
↪ X: *see* Figure 1.8
↪ X: *see* Figure 1.9
↪ Radioisotopes (^{35}S, ^{33}P) present in position α (*see* Figure 1.5) for labeling by 3' extension
↪ Stored at –80°C or 4°C in aliquots
↪ Specific activity ≈ 3,000 Ci / mmol
↪ Treated with DEPC (*see* Appendix B1.2)
↪ *See* Appendix B2.12
↪ For labeling
↪ 1 M potassium cacodylate; 125 mM Tris–HCl; BSA 1.25 mg / L; pH 6.6
↪ Storage at –20°C
↪ 200 mM potassium cacodylate; 1 mM EDTA; 200 mM KCl; 0.2 mg BSA; 50% glycerol ($^v/_v$); pH 6.5
↪ Storage at –20°C (the presence of glycerol keeps the enzyme solution liquid at –20°C, and allows the pipetting of the enzyme without warming the whole tube; the enzyme is labile)
↪ Labeled oligonucleotides:
- $\alpha^{35}S$ – dATP (*see* Figure 1.5)
- $\alpha^{33}P$ – dATP (*see* Figure 1.5)
- ^3H – dATP (*see* Figure 1.4)

↪ Eppendorf type

1.3.5.4 Protocol for radioactive probes

1. ***Reaction mixture*** — Place the following reagents in a sterile tube in the order indicated:

• Sterile water	5 µL
• Oligonucleotides, 10 or 100 pmol	5 µL
• Reaction buffer 5X	**4 µL**
• 25 mM CoCl$_2$	**4 µL**
• 10 pmol labeled dATP	1 µL
• TdT 50 U/µL	1 µL

2. ***Incubation***   60 min at 37°C

3. ***Stopping the reaction*** — Add:
- EDTA 10 mM 2 µL

Next step — Precipitation with ethanol.

↪ For a final volume of 20 µL

↪ According to requirements

↪ If a red precipitate forms, should check the origin of the radioactive nucleotide
↪ ≈ 50 µCi in the case of the $\alpha^{35}S$ – dATP probe
↪ Labile enzyme; must not be brought up to room temperature
↪ Labeling carried out at 37°C (minimum 40 min); not necessary to prolong the incubation time, or to increase the amount of enzyme, because the 3' extension is generally very long (*see* specific activity)

↪ Optional step which can be replaced by a water bath at 4°C, or a heater (10 min at 65°C)
↪ *See* Section 1.4.1.

1.3.5.5 Protocol for antigenic probes

1. *Reaction mixture* — Place the following reagents in a sterile tube, in the order indicated:

• Sterile water	4 µL
• Oligonucleotides, 10 or 100 pmol	5 µL
• Reaction buffer 5X	4 µL
• 25 mM CoCl$_2$	4 µL
• 1 mM labeled dUTP	1 µL
• 10 mM dATP	1 µL
• TdT 50 U / µL	1 µL

2. *Incubation*   60 min at 37°C

3. *Stopping the reaction* — Add:

• 10 mM EDTA   2 µL

Next step — Precipitation.

↪ Labeled oligonucleotides: (biotin, digoxigenin, fluorescein) – dUTP.
↪ Use Eppendorf type
↪ Add for a final volume of 20 µL.
↪ The oligonucleotides are diluted in sterile water and stored at –20°C.

↪ Only 1 to 2 labeled nucleotides are added at the 3' end.
↪ The addition of unlabeled dATP allows the incorporation of several labeled dUTPs.
↪ The enzyme deoxynucleotide terminal transferase catalyzes the polymerization of nucleotides from the 3' end.
↪ This is a labile reagent and must not be reheated.
↪ It is not necessary to prolong the incubation time, or to increase the amount of enzyme, as the extension remains limited.
↪ This is an optional step which can be replaced by a water bath at 4°C or a heater (10 min at 65°C).

↪ The oligonucleotides are purified by ethanol precipitation (*see* Section 1.4.1).

1.3.6 5' Extension

The labeling by 5' extension of an oligonucleotide consists of phosphorylating the 5' end.

1.3.6.1 Principle

This additional reaction is carried out by polynucleotide kinase (an enzyme extracted from the calf thymus).

❑ *Advantages*
• Only small quantities of oligonucleotides required
• Specific activity

• Quantification

❑ *Disadvantages*
• Low specific activity

• Poor resolution

• Cannot be used with antigenic labels

↪ The phosphate concerned is situated in position γ of a radioactive nucleotide (*see* Figure 1.5).

↪ This is an enzymatic reaction.
↪ The labeling obtained is limited to a single phosphate group, and is thus very weak.
↪ It is possible to label both single- and double-stranded DNA.
↪ It allows the determination of the yield of the labeling.
↪ Accurate estimation of the signal is possible: one hybrid = one isotope.

↪ A single phosphate is added to the oligonucleotide.
↪ The addition of an isotope of phosphorus does not give good resolution.
↪ It is necessary to use labels located on the phosphate in position γ (*see* Figure 1.5).

1.3.6.2 Summary of the different steps

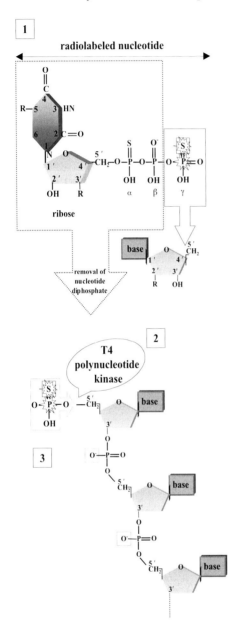

1 = Radioactive nucleotide of high energy (^{35}S or ^{32}P) — Labeling of the phosphate group in position γ.

2 = Enzymatic action.

3 = Labeling by a group containing a single atom of phosphorus or sulfur.

Figure 1.17 Labeling by 5' extension.

1.3.6.3 Equipment/reagents/solutions

1. Equipment
- Water bath / incubator at **37°C**
- Vortex mixer

⇝ Enzymatic reaction

1.3 Labeling Techniques

2. Reagents

- Labeled ATP

- DTT
- EDTA
- Imidazole
- MgCl$_2$
- T4 polynucleotide kinase
- Spermidine
- Tris

⇝ Molecular biology grade
⇝ Only to be used for *in situ* hybridization
⇝ Label required to be carried by the phosphate in position γ

⇝ Labile enzyme

3. Solutions

- Sterile water
- Phosphorylation buffer 10X (500 mM Tris – HCl; 100 mM MgCl$_2$; 1 mM EDTA; 50 mM DTT; 1 mM spermidine; pH 8.2)

⇝ **All the solutions are prepared using DNase - free reagents in a sterile container** (*see* Appendix A2).
⇝ *See* Appendix B1.1.
⇝ Stored at room temperature after sterilization

1.3.6.4 Protocol for radioactive probes

⇝ Oligonucleotides labeled by γ (^{32}P/^{33}P or ^{35}S) – ATP

1. Phosphorylation — Dissolve:

• DNA	25 pmol
• Phosphorylation buffer 10X	5 µL
• Labeled ATP	1 µL
• Polynucleotide kinase 50 U / µL	1 µL

⇝ 25 µg DNA

⇝ Catalyzes the polymerization of nucleotides from the 5' end
⇝ Labile reagent; must not be brought to room temperature

2. Incubation   60 min at 37°C

⇝ No necessity to extend the time

3. *Stopping the reaction* —
    Add 10 mM EDTA   2 µL

⇝ Optional step which can be replaced by a water bath at 4°C or a heater (10 min at 65°C).

Next step — Precipitation.

⇝ The oligonucleotides are purified by ethanol precipitation (*see* Section 1.4.1).

1.3.7 Addition

Chemically add antigenic labels to a molecule of DNA.

⇝ Reaction possible with synthetic oligonucleotides

1.3.7.1. Principle

A photoactive molecule has the property of creating a free covalent bond by the effect of photons.

⇝ There are other types of molecules which can interact with DNA (e.g., ethidium bromide, DAPI, DIPI).

In the presence of this molecule and of DNA, an additive reaction occurs between the photoactive antigen and the nucleic acid by activation due to the photons.

❏ *Advantage*
- Addition achieved on double-stranded DNA

❏ *Disadvantages*
- Yield of the addition reaction weak
- Large amount of DNA required
- Labile reagents

↪ It is necessary to separate the different constituents after the reaction.

↪ Addition of the antigen does not change the capacity of hybridization of this DNA.

↪ Low specific activity
↪ ≈ 1 µg
↪ Storage in the cold and away from light

1.3.7.2 Summary of the different steps

1 = **Incubation** of the photoactive reagent with double-stranded DNA

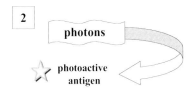

2 = **Illumination** of the photoactive reagent solution containing double-stranded DNA.

3 = **Precipitation** of the DNA.

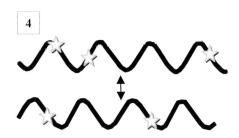

4 = **Denaturation**.

Figure 1.18 Labeling by addition.

1.3.7.3 Equipment/reagents/solutions

1. *Equipment*
- Ice bath
- Water bath
- Mercury-vapor lamp

↪ Enzymatic reaction

↪ Characteristics depend on the photoactive reagent

2. Reagents

- Ethanol 100%
- 2-Butanol
- Photoactive antigen

- EDTA
- NaCl
- Tris

3. Solutions

- Sterile water
- 100 mM EDTA; pH 8.0
- Ethanol 100%
- 4 M NaCl
- 100 mM Tris – HCl buffer; pH 9.5

↪ **Molecular biology grade**
↪ Only used for *in situ* hybridization

↪ All the antigenic labels for *in situ* hybridization are available in this form.

↪ **All the solutions are prepared using DNase - free reagents in a sterile container** (*see* Appendix A2).
↪ DEPC-treated (*see* Appendix B1.2)
↪ *See* Appendix B2.12
↪ Stored at –20°C
↪ *See* Appendix B2.8
↪ *See* Appendix B3.6.1

1.3.7.4 Protocol

1. Place the following reagents in a sterile Eppendorf tube (on ice), in the indicated order:
   - DNA

   - Photoprobe

2. Irradiate the mixture — Mercury-vapor lamp (distance between the reaction mixture and the lamp: 10 cm). **15 min**

3. Complete the reaction volume — Tris – HCl buffer. **to 100 µL**

4. Add and mix 2-butanol. **100 µL**

5. Centrifuge. **1200 g for 5 min**

6. Leave out the alcohol phase.

7. Repeat the extraction by adding 2-butanol. **100 µL**

8. Add:
   - NaCl **75 µL**
   - Ethanol **100 µL**

Next step — Purification by precipitation.

↪ The proportions of the two constituents depend on the length of the DNA

↪ DNA   > 100 bp   1:1 ($^v/_v$)
↪ DNA   > 50 - 100 bp   2:1 ($^v/_v$)
↪ DNA   > 20 - 50 bp   3:1 ($^v/_v$)

↪ The efficiency of the lamp decreases in proportion to time of usage.
↪ It is possible to replace irradiation by an incubation at 90°C for 30 min.

↪ *See* Section 1.4.1. This labeled probe can be stored for more than a year at –20°C.

1.3.8 *In Vitro* Transcription

1.3.8.1 Principle

RNA probes are obtained from an *in vitro* transcription system, and are labeled during synthesis.

The transcription vector (plasmid) into which the DNA is inserted must possess the RNA promoters situated on both sides of the cloning site.

For transcription, the plasmid is linearized by cutting between the insert and the promoter not in use.

Depending on which promoter is used, transcription will produce sense or anti-sense probes.

Transcription begins at the transcription initiation site (the site of insertion of the promoter). The growing strand extends in the 5' → 3' direction. Each nucleotide put into place is complementary to the corresponding nucleotide in the insert.

Transcription can be performed according to the polymerase used, so that either sense mRNA (sense probe) or anti-sense probes (from the opposite strand) can be produced.

• Transcription with polymerase 1

• Transcription with polymerase 2

When the enzyme arrives at the site of cutting, it uncouples from the template and retranscribes the insert.

❏ *Advantages*
• An RNA probe is produce.
• The probes obtained are single stranded.
• The amount of labeled probe is greater than the amount of template.

↪ Transcribes DNA onto an RNA probe

↪ The most frequently used promoters for RNA polymerase are SP6, T3, and T7.

↪ This step presupposes that the orientation of the cDNA in the vector can be read from the restriction map.
↪ Labeling is achieved by the action of RNA polymerase, which incorporates successive labeled or unlabeled nucleotides during the course of RNA synthesis.
↪ The newly synthesized RNA is labeled.

↪ The single-stranded RNA formed is complementary to the DNA strand which serves as its template.

↪ Polymerase 1 makes labeled copies of the same sequence as the target nucleic acid. The synthesis of the RNA strand serves as a control for determining background labeling (no specific hybridization with the target RNA).
↪ Polymerase 2 makes labeled copies complementary to the target nucleic acid (hybridization with the target RNA). This synthesis of the anti-sense RNA is the probe.
↪ Depending on the stability of the enzymes and the strength of the promoter, several cycles of initiation on the same DNA template can considerably amplify the amount of RNA transcribed, in relation to the DNA template.
↪ The RNA obtained is single stranded.

↪ RNA – RNA hybrids the most stable

↪ Synthesis of labeled transcripts (amplification 10 X with labeled molecules)

- The specific activity is high.

- The probe is large.
- It is possible to obtain both sense and anti-sense probes

❑ *Disadvantages*
- Insertion of the cDNA into a vector to obtain transcription
- Lability of the probes

- Necessitates the use of a restriction enzyme to linearize the plasmid
- Necessitates cutting the probes to obtain small fragments

⇝ Possibility of using several labeled nucleotides
⇝ Copies the whole insert

⇝ Construction is very important.

⇝ This avoids all risk of contamination by RNases.
⇝ Restriction enzymes have a limited life span, and are, moreover, very labile.
⇝ These probes are of equivalent length to that of the DNA which serves as template.

1.3.8.2 Summary of the different steps

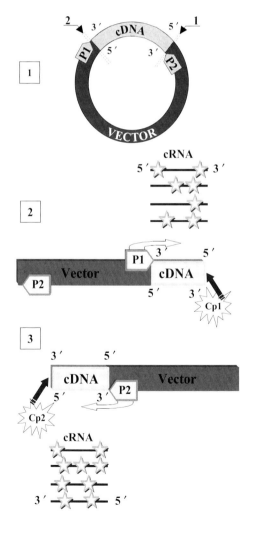

1 = Insertion into a plasmid:
- The target cDNA is inserted into a plasmid equipped with two promoters, P1 and P2, recognized by two different RNA polymerases and situated on either side and on each strand of the insert.
- Two cutting sites are present between the insert and each promoter. They make possible the linearization of the plasmid by means of a cut between the insert and the promoter not in use.

P1 = promoter 1
P2 = promoter 2

2 / 3 = Creation of RNA probes by *in vitro* transcription in the presence of labeled nucleotides.

2 = Creation of sense RNA probes — The sense transcripts are obtained by the action of the polymerase attaching itself to the promoter P1 (sense probe) (i.e., a copy of the target sequence).

3 = Creation of complementary RNA probes — The anti-sense transcripts are obtained by the action of the polymerase attaching itself to the promoter P2 (anti-sense probe) (i.e., complementary to the target sequence).

Figure 1.19 *In vitro* transcription.

1.3.8.3 Equipment/reagents/solutions

1. *Equipment*
- Water bath at **37°C**
- Centrifuge
- Vacuum jar or Speedvac
- Electrophoresis tank
- Power pack
- Vortex mixer

↪ Enzyme action
↪ ≥14000 g
↪ To dry the probe

2. *Reagents*

↪ Molecular biology grade, RNase-free, only to be used for labeling

- Agarose
- Isoamyl alcohol
- Transfer RNA
- Chloroform
- Cobalt chloride
- EDTA
- Ethanol 100%
- Phenol
- Plasmid containing the insert to be transcribed
- Antigenic ribonucleotides
  — Biotin – X – UTP
  — Digoxigenin – X – UTP
  — Fluorescein – X – UTP
- Radioactive ribonucleotides
- Tris

↪ *See* Figure 1.3.
↪ *See* Figure 1.7.
↪ *See* Figure 1.8.
↪ *See* Figure 1.9.
↪ Labeled in position α (*see* Figure 1.5)

3. *Solutions*

↪ **All the solutions are prepared using RNase-free reagents in a sterile container** (*see* Appendix A2).

- Transfer RNA (tRNA) 10 mg / mL
- Sterile water
- 500 mM EDTA; pH 8.0
- Ethanol 70%, 100%
- Transcription kit
- 10 mM ribonucleotides (rNTP mix) ATP, GTP, CTP

- Antigenic ribonucleotides
- Radioactive ribonucleotides
- RNA polymerases

- DNase I
- 10% SDS
- Chloroform / isoamyl alcohol
  — Isoamyl alcohol **1 vol**
  — Chloroform **49 vol**

↪ *See* Appendix B2.4.
↪ DEPC-treated (*see* Appendix B1.2).
↪ *See* Appendix B2.12.

↪ Labeling kit; storage at –20°C
↪ Each nucleotide diluted in water; rNTP mix neutralized at pH 7.0
↪ Storage at –20°C
↪ Storage at –20°C
↪ Storage at –80°C
↪ To be determined according to the promoter used, which is generally T3, T7, or SP6
↪ *See* Appendix B2.11.

1.3 Labeling Techniques

- Buffers
  — Transcription buffer 10X

  ⇝ Labeling kit
  ⇝ 400 mM Tris – HCl; 60 mM MgCl$_2$; 100 mM DTT; 20 mM spermidine; pH 8.0 (all present in the labeling kit)

  — TBE buffer 5X

  ⇝ 450 mM Tris; 450 mM boric acid; 10 mM EDTA

  — TE buffer 10X

  ⇝ See Appendix B3.5 (100 mM Tris; 10 mM EDTA)

1.3.8.4 Protocol for radioactive probes

⇝ The wearing of gloves is essential when handling riboprobes (to avoid degradation of the RNA by exogenous RNases).

1. Linearization of the plasmid

- Linearized plasmid   ≥100 µL
  (≈ 20 to 50 µg)

  ⇝ The plasmid must first be precipitated, then purified on a gel and finally linearized. It is taken up in a minimum volume.

- Denature   5 to 10 min at 92°C

2. Purification

a. Add phenol / chloroform.   v/v

⇝ Vortex for 1 min (formation of an emulsion)

b. Centrifuge.   ≥ 14,000 g for 5 min

⇝ To recover the supernatant (the DNA is in the upper aqueous phase)

3. Precipitation

⇝ See Section 1.4.

4. In vitro transcription

⇝ Riboprobe labeled with ^{35}S

a. Place the following reagents in a sterile Eppendorf tube, on ice, in the indicated order:

- Linearized plasmid (1 µg / µL)   1 µL
- Labeled nucleotide   2 µL
- Sterile H$_2$O   to 20 µL
- Transcription buffer 10X   2 µL
- RNasin 20 U/µL   1 µL

⇝ Can be replaced by PCR fragments containing a promoter
⇝ Final concentration (α^{35}S – UTP / 1 µM)
⇝ See Appendix B1.1.

⇝ Final concentration 1 U/µL
⇝ Inhibitor of RNases, to avoid the degradation of synthesized RNA. This reagent does not withstand high temperatures. Add at the last moment.

- NTP mix   2 µL
  — 2.5 mM ATP
  — 2.5 mM GTP
  — 2.5 mM CTP
- RNA polymerase 20 U/µL   2 µL

⇝ Final concentration 0.4 mM
⇝ First, prepare a mixture of all the constituents, apart from DNA and RNA polymerase, and neutralize at pH 7.0. Add labeled UTP.
⇝ The polymerases (SP6, T3, or T7) allow the synthesis of the anti-sense and sense riboprobes. The latter serves as a control for the determination of background labeling.

Probes

b. Incubate.	**60 min at 37°C**	↪ Do not extend the incubation time, to avoid the degradation of the RNA.
5. *Digestion of the DNA*		
a. Add molecular biology grade DNase I 1 µg/µL.	**2 µL**	
b. Incubate.	**15 min at 37°C**	↪ To eliminate specifically the strands of DNA (elimination of the template)
c. Add 100 m*M* EDTA.	**2 µL**	↪ Enzyme inhibited by EDTA (final concentration: 4 m*M*).
6. *Extraction of proteins*		↪ Extraction with phenol, and precipitation in alcohol
a. Add tRNA 10 mg / mL.	**1 µL**	↪ Final concentration: 400 µg / mL
b. Denature.	**5 min at 95°C**	↪ Or 2 min at 100°C
c. Add • 10% SDS • TE	**4 µL** **to 100 µL**	
• Phenol (1 vol)	**100 µL**	↪ Vortex 1 min
d. Centrifuge.	**≥ 14,000 g for 5 min**	↪ Extract the upper aqueous phase
• Chloroform / isoamyl alcohol	**100 µL**	↪ Chloroform - isoamyl alcohol (49:1) ($^v/_v$)
e. Centrifuge.	**≥ 4000 g for 10 min**	

Next step — Purification, precipitation.   ↪ *See* Section 1.4.

1.3.8.5 Protocol for antigenic probes

The reaction conditions are identical to those described for radioactive riboprobes.

↪ **The wearing of gloves is essential when handling riboprobes (to avoid degradation of the RNA by exogenous RNases)**

Steps 1, 2, and 3 are identical to those for labeling using radioactive isotopes.

↪ *See* Section 1.3.8.4.

4. *In vitro transcription*

a. Place the following reagents in a sterile Eppendorf tube, on ice, in the indicated order: • Linearized plasmid (1 µg/µL)	**1 µL**	↪ It is possible to replace by PCR fragments containing a promoter.

• Labeled nucleotide	2 µL	↝ Use UTP–(biotin, digoxigenin, fluorescein) (3.5 mM)
• Sterile H$_2$O	to 20 µL	↝ *See* Appendix B1.1.
• Transcription buffer 10X	2 µL	
• RNasin 20 U/µL	**1 µL**	↝ Add to a final concentration 1 U / µL. ↝ Inhibitor of RNases to avoid the degradation of synthesized RNA — this reagent does not withstand high temperatures. Add at the last moment.
• NTP mix — 10 mM ATP — 10 mM GTP — 10 mM CTP — 6.5 mM UTP	2 µL	↝ Add to final concentration 0.4 mM. ↝ First, prepare a mixture of all the constituents, apart from the DNA and RNA polymerase, and neutralize at pH 7.0. Add labeled UTP.
• RNA polymerase 20 U/µL	2 µL	↝ The polymerases (SP6, T3, or T7) allow the synthesis of the anti-sense and sense riboprobes. The latter serves as a control for the determination of background labeling.
b. Mix and centrifuge.		↝ All the constituents must be at the bottom of the tube.
c. Incubate.	**2 h at 37°C**	↝ Do not extend the incubation time, to avoid the degradation of RNA.

Steps 5 and 6 are identical to those for transcription using a radiolabeled nucleotide. ↝ *See* Section 1.3.8.4.

Next step — Purification, precipitation. ↝ *See* Section 1.4.

1.4 PRECIPITATION TECHNIQUES

Purification of DNA can be achieved:

- By ethanol precipitation ↝ The labeled probe is precipitated in ethanol to remove the free nucleotides and to concentrate the probe.

- On a Sephadex G-50 column ↝ Only the nucleotides are removed. If the volume is very large, it is necessary to precipitate the probe.

1.4.1 Ethanol Precipitation

1.4.1.1 Principle

Nucleic acids are water-soluble molecules that precipitate in an alcoholic solution in the presence of salts.

1.4.1.2 Equipment/reagents/solutions

1. *Equipment*
   - Centrifuge
   - Vacuum jar or Speedvac
   - Freezer –20 or –80°C

   - Vortex mixer

   ↪ ≥14,000 g
   ↪ To dry the probe
   ↪ Storage of reagents and precipitation products

2. *Reagents*

   ↪ Molecular biology grade, only to be used for precipitation

   - Ammonium acetate
   - Sodium acetate
   - Transfer RNA
   - Ethanol 100%

3. *Solutions*

   ↪ **All the solutions are prepared using RNase and DNase-free reagents in a sterile container** (*see* Appendix A2).

   - 3 M sodium acetate; pH 5.2
   - 7.5 M ammonium acetate; pH 5.5
   - Transfer RNA (tRNA) 10 mg / mL
   - 4 M lithium chloride
   - Ethanol 100% and 70%

   ↪ *See* Appendix B2.2
   ↪ *See* Appendix B2.1
   ↪ *See* Appendix B2.4
   ↪ Stored at –20°C
   ↪ Stored at –20°C

1.4.1.3 Protocol

1. *Reaction mixture*
   a. On ice, add the reagents in the following order:
   - tRNA 10 mg / mL   **2 µL**

   ↪ This facilitates the precipitation of oligonucleotide probes. **Optional.**

   - 7.5 M ammonium acetate,   **1 / 5 of the final volume**
   - 3 M sodium acetate, or
   - 4 M lithium chloride
   - Ethanol 100%   ≈ **2 to 3 vol**

   ↪ Ammonium acetate can be replaced by 3 M sodium acetate or 4 M lithium chloride. In the latter case, it is necessary to wash the pellet with 70% alcohol (stored at –20°C) after precipitation and recentrifugation.
   ↪ Store ethanol at –20°C.
   ↪ Volume of the reaction solution

   b. Vortex, centrifuge.

   ↪ No reagent must remain on the side of the tube.

2. *Incubation*
   a. Incubate.   **60 min at –80°C or overnight at –20°C**

   ↪ Precipitation of the probe
   ↪ Minimum 30 min at –80°C or
   ↪ Minimum 2 h at –20°C

1.4 Precipitation Techniques

    b. Centrifuge.   > 14,000 g for ≥ 15 min at 4°C

↝ Orient the tube to find out the position of the pellet and remove the supernatant more easily. The accelerating power of the centrifuge is essential in obtaining a pellet. In the case of labeling using a radioactive nucleotide, the supernatant contains all the free radioactive dNTP. It must be removed.

3. Washing
    a. Remove the supernatant.
    b. Wash the precipitate with ethanol 70%   50 µL

↝ Elimination of salts

    c. Centrifuge.   > 14000 g for ≥ 15 min at 4°C

↝ Orient the tube to facilitate the observation of the position of the pellet and the removal of the supernatant.

    d. Remove the supernatant.
    e. Dry.

1.4.2 Passage through a Column

1.4.2.1 Principle

The nucleic acids are separated from the free nucleotides and labeling reagents by passage through a column.

1.4.2.2 Equipment/solutions
1. *Equipment*
   - Centrifuge
   - Vacuum jar or Speedvac
   - Sephadex G-50 column
   - Vortex mixer
2. *Solutions*

   - TE buffer

↝ ≥14,000 g
↝ To dry the probe

↝ **Solutions are prepared using RNase- and DNase-free reagents in a sterile container** (*see* Appendix A2).
↝ *See* Appendix B3.5.

1.4.2.3 Protocol
1. Wash the column with the buffer.
2. Drop on the labeled solution.
3. Separate the constituents.

↝ By pressure ("Push column") or centrifugation

4. Wash the column.
5. Elute the probe.
6. Precipitate the probe if necessary.

↝ *See* Section 1.4.1.

1.4.3 Drying

↝ Eliminates all trace of volatile alcohol

The pellet is dried by:
- Vacuum jar or   30 min

- Speedvac 5 to 10 min

↪ The probes can be stored dry or solubilized in a buffer (e.g., TE, *see* Appendix B3.5), or in sterile water, and stored at –20°C.

1.5 CONTROLS FOR LABELING

↪ An additional (but very important) step

1.5.1 Radioactive Probes

1.5.1.1 Calculating the incorporated radioactivity
1. Add:
   - TE 10X or DEPC-treated water 80 µL
   - DTT 100 mM 20 µL

↪ *See* Appendices B1.2 and B 3.5

↪ Final concentration: 20 mM
↪ Radioactivity incorporated into the probe / radioactivity due to the nucleotide introduced into the labeling tube.
↪ If the percentage of incorporation is ≤50 %, the enzyme activity is too weak, or the nucleotide is degraded.

2. Check the incorporation of radioactive nucleotides by counting in scintillation fluid.
   - Original solution 1 µL

1.5.1.2 Calculation of specific activity

Radioactivity incorporated / mass of the probe or molarity of the probe

↪ The specific activity of a probe is the best indication of labeling reproducibility, and allows a comparison of activities of different probe types: cDNA, cRNA, or oligonucleotide. Too high specific activity is not always the guarantee of a good *in situ* hybridization signal. In the case of an oligonucleotide, a high specific activity corresponds to a 3' extension much longer than the oligonucleotide itself. In the case of cDNA or cRNA, it may lead to rapid radiolysis and fragmentation of the probes.

- Labeled probe 1 µL
- Scintillation fluid > 1 mL

1.5.1.3 Verification of labeling on an analytical gel of alkaline agarose (for probes labeled by nick translation) or agarose / polyacrylamide (used for cRNA / oligonucleotide probes).

↪ Verification of the labeling and of the size of the fragments obtained

1.5 Controls For Labeling

1. **Equipment**
   - Autoradiographic cassette
   - Electrophoresis tank
   - Electrophoresis power pack
   - Cellophane paper
   - Gel dryer

2. **Solutions**
   - Alkaline agarose buffer (alkaline agarose gels) ⇒ 50 mM NaOH; 1 mM EDTA
   - TBE buffer 5X (agarose / polyacrylamide gels) ⇒ 450 mM Tris; 450 mM boric acid; 10 mM EDTA

3. **Protocol**
   a. Deposit sample.   0.5 µL / dot   ⇒ About 500,000 cpm / well
   b. Allow to migrate (5 to 7.5 cm).   30 to 45 min at 90 V   ⇒ Necessary for the sample and electrophoresis buffers to be freshly prepared and sterilized
   c. Dry the gel.   ⇒ Gel dryer on Whatman paper
   d. Expose the gel covered in cellophane to autoradiographic film.   6 to 8 h   ⇒ Kodak XAR film
   ⇒ Visualization of the bands corresponding to the radioactive probes (exposure of the film)

1.5.2 Antigenic Probes

1.5.2.1 Controls

Labeling is achieved by creating a range of dilutions on a nylon membrane. After drying, the labeling of the probe is visualized (Figure 1.20).
- Dilutions by a factor of 10: (1, 1:10, 1:100, 1:1000)

⇒ A labeling standard is provided in certain kits.

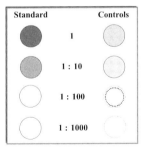

Figure 1.20 Controls for antigenic probes.

1.5.2.2 Biotin and digoxigenin detection

If the label is biotin or digoxigenin, this is revealed by immunoenzymatic detection of the antigenic molecule.

⇒ The anti-digoxigenin or anti-biotin antibodies used can be conjugated to alkaline phosphatase. In this case, the chromogen used is NBT–BCIP (Nitroblue tetrazolium / 5-bromo-4-chloro-3-indolyl phosphate) (*see* Section 5.2.3).

Probes

1.5.2.3 Visualization of fluorescein

If the label is fluorescein or any other fluorescent label, it is possible to place an aliquot on nylon membrane or filter paper and observe the fluorescence under a microscope. A signal is evidence of probe labeling.

↬ It is sometimes necessary to wash the membrane to remove unincorporated nucleotides if precipitation has not been carried out.
↬ This control is quick, and can be carried out before the precipitation of the probe.

1.5.2.4 Disadvantages

- Uses a relatively large amount of probe (up to 10 ng).
- Does not allow the determination of the specific activity of the probe.

1.6 STORAGE / USE

↬ Using a probe at the time of preparation removes all the risks inherent in storage.

1.6.1 Storage

Storage depends upon the nature of the probe and label:

1. Probes
- Antigenic probe

- Radioactive probe

↬ DNA probes (cDNA and oligonucleotides) keep for much longer than cRNA probes.
↬ Radioactive labels lead to radiolysis of nucleic acids.

2. Labels
- Radioactive label

↬ Storage is dependent on half-life:

Isotope	Half-life	Storage
^{33}P	25.4 days	1 week
^{35}S	87.4 days	up to 1 month
^{3}H	12.4 years	up to 1 year
^{32}P	14.3 days	

- Antigenic label

↬ Antigenic labels are very stable. When dissolved in TE buffer at $-20°C$ they are generally stable for several months.

Storage methods
- Dry

- In sterile water or buffer

- In a hybridization buffer

↬ At the moment of use, take up the desired amount of probe in hybridization buffer.
↬ Storage is possible at $-20°C$ for several months.
↬ Storage is possible at $-20°C$ for several months.

1.6.2 Use

In the case of double-stranded probes, DNA principally, it is necessary to denature the probe:

1. Denature. **10 min at 92°C or 5 min at 96°C**

2. Place on ice.

3. Put the probe on the sections immediately.

↪ Certain long RNAs must also be denatured (*see* Chapter 3). This step is carried out in a hybridization buffer.

↪ If a probe is labeled with ^{35}S, DTT (10 mM) must be added after denaturation, since the probe will be heat sensitive (Figure 1.21).

↪ Prevents the two complementary strands from rehybridizing.

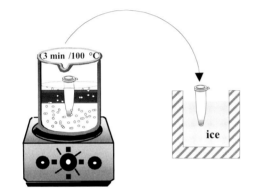

Next step — Hybridization.

Figure 1.21 Denaturation of the probe.

↪ *See* Chapter 4

Chapter 2

Tissue Preparation

Contents

2.1	Tissue Removal		65
	2.1.1	Tissue Origin	65
	2.1.2	Precautions	66
2.2	Fixation		66
	2.2.1	Principle	66
		2.2.1.1 Fixation Conditions	67
		2.2.1.2 Fixatives	67
		2.2.1.3 Choice of Protocol	68
	2.2.2	Types of Fixative	68
		2.2.2.1 Cross-Linking Fixatives	68
		2.2.2.2 Precipitating Fixatives	69
		2.2.2.3 Mixtures	70
	2.2.3	Equipment / Solutions	70
		2.2.3.1 Equipment	70
		2.2.3.1.1 Fixation by Perfusion	70
		2.2.3.1.2 Fixation by Immersion	70
		2.2.3.1.3 Cells in Suspension	70
		2.2.3.2 Solutions	70
	2.2.4	Fixation Protocols	71
		2.2.4.1 Tissues	71
		2.2.4.2 Cells in Suspension	72
		2.2.4.3 Cells in Culture/Smears	73
2.3	Frozen Tissue		74
	2.3.1	Summary of the Different Steps	74
	2.3.2	Equipment/Reagents/Solutions	75
		2.3.2.1 Equipment	75
		2.3.2.2 Small Equipment	75
		2.3.2.3 Pretreated Equipment	75
		2.3.2.4 Reagents	75
		2.3.2.5 Solutions	75
	2.3.3	Cryoprotection	76
		2.3.3.1 Principle	76
		2.3.3.2 Cryoprotectants	76
		2.3.3.3 Method	76
	2.3.4	Freezing	76
		2.3.4.1 Principle	76
		2.3.4.2 Freezing Agents	76
		2.3.4.3 Method	77
		2.3.4.4 Storage	78
	2.3.5	Preparation of Frozen Sections	78
		2.3.5.1 Equipment	78
		2.3.5.2 Protocol	78
	2.3.6	Advantages/Disadvantages	81

2.4　Tissue Embedded in Paraffin Wax 81
    2.4.1　Summary of the Different Steps 82
    2.4.2　Equipment/Reagents/Solutions 83
        2.4.2.1　Equipment 83
        2.4.2.2　Small Equipment 83
        2.4.2.3　Pretreated Equipment 83
        2.4.2.4　Reagents 83
        2.4.2.5　Solutions 83
    2.4.3　Embedding in Paraffin Wax 84
        2.4.3.1　Aim 84
        2.4.3.2　Protocol 84
    2.4.4　Preparation of Paraffin Sections 85
        2.4.4.1　Equipment 85
        2.4.4.2　Protocol 86
    2.4.5　Dewaxing the Sections 87
        2.4.5.1　Aim 87
        2.4.5.2　Method 88
    2.4.6　Advantages/Disadvantages 88

The aim of this chapter is to present the procedures and precautions required to prepare a particular sample for *in situ* hybridization in the kind of conditions that will optimize the preservation of both the morphology and the structure of the nucleic acids.

2.1 TISSUE REMOVAL

⇝ This step is often the cause of negative or disappointing results.

2.1.1 Tissue Origin

Various types of biological material can be used:
- Whole organs
- Biopsies
- Cells in culture
  — in monoculture
  — in suspension

- Cellular smears

⇝ For histological sections or whole mounts
⇝ For sections
⇝ *See* Figure 2.1.
⇝ On a glass coverslip
⇝ In a culture bottle:
  • after detachment and cell centrifugation
  • after resuspending the pellet
⇝ Cells deposited on a glass slide using a cytocentrifuge (essential in the case of a weak suspension of cells)
⇝ Preparation obtained by spreading out the biological fluid: the technique of smearing allows the cells to be spread in a plane (in monocellular culture) on a glass slide

2.1.2 Precautions

The precautions are similar to those required to preserve sterility.

⇝ Contamination by external RNases or DNases must be kept to a minimum:
- Gloves must be worn.
- All instruments and slides must be sterilized (*see* Appendix A1.1).

1 = Cellular suspension:
- Cultured cells
- Dissociated cells
- Suspended cells

2 = Preparation of a cellular pellet:
- Centrifugation
- Fixation
- Freezing (*see* Section 2.3.4) or embedding (*see* Section 2.4.3)
- Cutting (*see* Sections 2.3.5 and 2.4.4)

3 = Preparation of a cellular spread:
A: Smears if the cellular concentration is high
B: Cell spins if the cellular concentration is low
- Fixation (*see* Section 2.2.4)

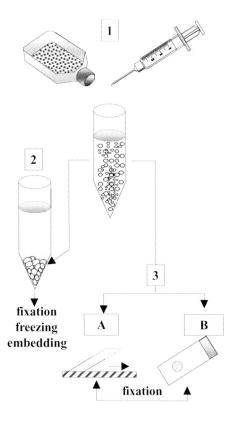

Figure 2.1 Cytological preparations.

2.2 FIXATION

2.2.1 Principle

The fixation process is essential to the success of *in situ* hybridization. It consists of allowing certain chemicals, called fixatives, to enter the aqueous cellular phase. The aim is to inhibit cellular metabolism, to deactivate lysosomal enzymes and endogenous RNases and, at the same time, to conserve cellular morphology and nucleic acid integrity.

⇝ The tissue must be in a physiological medium that makes it possible to maintain a morphological state as close as possible to the living state.
⇝ It is pointless to start a study if the tissue is poorly fixed.
⇝ The conservation of RNA is more critical than that of DNA.

2.2 Fixation

To have the best chance of success, fixation must take place as soon as possible:
- After tissue removal, or
- *In vivo*

↪ Fixation by immersion
↪ Fixation by perfusion

2.2.1.1 Fixation conditions

To obtain good-quality fixation, it is necessary to take account of the following variables:

- Temperature

  ↪ The work should be done at the lowest possible temperature to reduce to a minimum the activity of the enzymes that bring about cellular autolysis.

- pH

  ↪ The pH of the fixative can have a significant effect on the conservation of the nucleic acids, which is the reason relatively neutral buffers (pH 7.2 to 7.4) are used.

- Concentration of fixative

  ↪ A high concentration can lead to overfixation, which will limit access to the nucleic acids.

- Osmolarity

  ↪ The osmolarity of the fixative mixture (buffer + fixative) must correspond to the physiological osmotic pressure. The addition of sodium chloride or variable sucrose concentrations allows this osmotic pressure to be adjusted.

- Fixation time

  ↪ This varies according to the origin and size of the biological sample.

The results of *in situ* hybridization depend, to a large extent, on the choice of fixative.

↪ Polyvalent fixatives provide the best compromise between the preservation of morphology and that of nucleic acids.

There is now a large variety of fixatives and fixation protocols that can be used with the different types of biological material.

↪ Each fixative has its own particular properties, including advantages and disadvantages. There is no universal fixative.

2.2.1.2 Fixatives

↪ Find the best fixative for each technique.

1. Cross-linking fixatives

↪ *Fixations that cause a linkage between the different cellular and tissue structures, e.g., formaldehyde or glutaraldehyde (see Section 2.2.2.1)*

2. Precipitating fixatives

↪ Fixations that lead to the precipitation of proteins, e.g., alcohol or acetone (*see* Section 2.2.2.2)

3. Mixtures

↪ Intended to maximize advantages and minimize disadvantages

↪ In practice, however, should not be used for *in situ* hybridization, e.g., Bouin's, Zenker's, B-5 fixative (*see* Section 2.2.2.3)

2.2.1.3 Choice of protocol

Fixation can be done either before or after attachment to the slide:

1. **Fixation after removal** of the biological material. The sample will be:
   - Frozen
   - Embedded

2. **Fixation after spreading** the biological material on a histological slide

↪ The choice of fixation protocol depends on the origin of the biological material and the conditions of preparation.
↪ Organ, tissue, biopsy, cells in culture

↪ To obtain sections with a cryomicrotome
↪ To obtain sections of tissue embedded in paraffin
↪ Cell spins, cell smears, whole mounts, frozen tissue sections
↪ Necessary to pretreat the slides (*see* Appendix A3)

2.2.2 Types of Fixative

Fixatives are classified into two groups, according to their action on proteins:
- Cross-linking fixatives
- Precipitating fixatives

2.2.2.1 Cross-linking fixatives

1. **Type** — These are usually low-cross-linking aldehyde fixatives:
   - Formaldehyde (HCHO)

↪ Handle under a ventilated hood

↪ HCHO is a gas whose aqueous solution is commercial formol (37%), which always contains stabilizing additives (e.g., 10 to 15% methanol).
↪ Toxic if inhaled, or on contact with the skin or mucous membranes; handle under a ventilated hood.

- Paraformaldehyde

$$\underset{H}{\overset{H}{C}}=O + \underset{H}{\overset{H}{C}}=O \mid \underset{H}{\overset{H}{C}}=O \longrightarrow -\underset{H}{\overset{H}{C}}-O-\underset{H}{\overset{H}{C}}-O-\underset{H}{\overset{H}{C}}-$$

- Glutaraldehyde (CHO–(CH$_2$)$_3$–CHO)

2. **Properties** — These aldehyde fixatives are highly penetrating.

↪ Polymer of formaldehyde (*see* Figure 2.2); the advantage of paraformaldehyde over formol lies in the absence of stabilization additives, but it has to be prepared at the time of use.

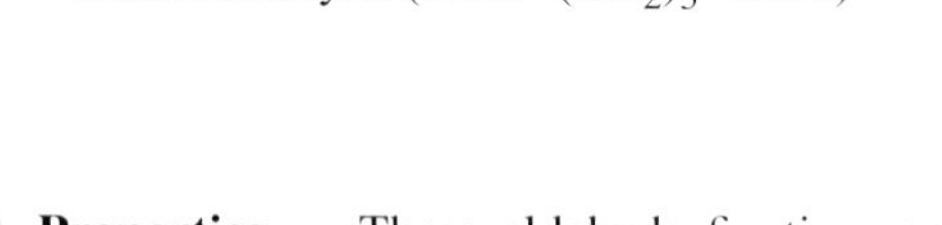

Figure 2.2 Polymerization of formaldehyde.

↪ Traces of this cross-linking fixative give good tissue preservation. It is used at concentrations < 0.5%.
↪ Toxic if inhaled.
↪ However, the best fixation is obtained with tissue fragments of small size (the speed of penetration is on the order of 1 mm / h).
↪ They remain the best compromise between the preservation of morphology and the *in situ* hybridization signal.

They act by the formation of methylene links, which are favored at pH 7.5.

❏ *Advantages*
- Rapid penetration
- The molecular link obtained allows the best accessibility of the probe to the target nucleic acid sequences
- Good fixation of proteins

❏ *Disadvantages*
- Rapid degradation after dilution

3. Preparation
- Formaldehyde
- 40% paraformaldehyde
- 4% paraformaldehyde
- Paraformaldehyde + glutaraldehyde

⇝ This is accomplished by using a sufficient quantity of fixative (a minimum of 10 times the sample volume) and a buffer solution.

⇝ Working concentration of 2 to 4%
⇝ Important for long probes

⇝ The most common fixative for *in situ* hybridization is paraformaldehyde.

⇝ Prepare 4% paraformaldehyde at the time of use.

⇝ *See* Appendix B4.3.
⇝ Stock solution (*see* Appendix B4.4.1)
⇝ Working solution (*see* Appendix B4.4.2)
⇝ Working solution (*see* Appendix B4.4.3)

2.2.2.2 Precipitating fixatives

1. **Type** — The fixing agents described as "precipitating," such as alcohol, acetic acid, and acetone, have been in use for a long time, and are sometimes still recommended:
- Alcohol (CH_3CH_2OH)

- Methanol (CH_3OH)

- Acetone (CH_3COCH_3)

2. **Properties** — Precipitating fixatives have the advantage of preserving nucleic acid structure, but the disadvantage of allowing nonspecific binding, which means that they produce background.
❏ *Advantages*
- Rapid and stable preparation
- Rapid penetration
- Preservation of nucleic acids
- Permeabilization

⇝ Precipitation of proteins is a means of arresting cellular metabolism and inhibiting the action of RNases and DNases.

⇝ Alcohol precipitates proteins and dissolves certain lipids, while preserving glycogen. It is used at an optimal concentration of 70 to 100% (generally 95%).
⇝ Alcohol is a very poor fixative for morphology in general.
⇝ Methanol is mainly used for frozen sections and cell spreads; generally used pure.
⇝ Acetone is mainly used for frozen sections and cell spreads. It can be used immediately after methanol, it is generally used pure.

⇝ The precipitation of proteins can reveal polar groups, which can cause adsorptions.

Tissue Preparation

❏ *Disadvantages*
- Poor morphological preservation

3. Preparation
- Acetone
- Alcohol
- Methanol

↝ *See* Appendix B4.1.
↝ *See* Appendix B4.2.1.
↝ *See* Appendix B4.2.2.

2.2.2.3 Mixtures

- B-5 fixative (formaldehyde, mercuric chloride, sodium acetate)
- Carnoy's fixative (ethanol, chloroform, acetic acid)
- Bouin's fixative (formol, picric acid, acetic acid)
- Zenker's fixative (acetic acid, mercury salts, potassium dichromate, disodium sulfate)

↝ These mixtures are not recommended for *in situ* hybridization:
- The acids they contain can hydrolyze nucleic acids, and thereby reduce the sensitivity and specificity of hybridization
- Mercury salts can form complexes with nucleic acids, thus limiting hybridization

2.2.3 Equipment / Solutions

2.2.3.1 Equipment

2.2.3.1.1 FIXATION BY PERFUSION
- Dissection scissors, metallic cannula, vascular clamp
- Peristaltic pump with adjustable flow

2.2.3.1.2 FIXATION BY IMMERSION
- Containers and caps
- Propipette

- Scalpel

2.2.3.1.3 CELLS IN SUSPENSION
- Cytocentrifuge

- Filters for the cytocentrifuge
- Pretreated slides
- Vortex mixer

↝ **All the equipment in contact with the tissue must be sterile.**

↝ This can be replaced by a suspended container.

↝ To fix the samples
↝ To withdraw toxic solutions without direct aspiration
↝ To cut the tissue into small fragments before fixation

↝ No special features, but the reservoirs must be sterile
↝ Sterile
↝ *See* Appendix A3
↝ Homogenization of biological liquids

2.2.3.2 Solutions
- 4% paraformaldehyde
- Phosphate buffer

↝ *See* Appendix B4.4.2.
↝ *See* Appendix B3.3. Cacodylate buffer is recommended for vegetable tissue to avoid precipitation due to the phosphate.

2.2.4 Fixation Protocols

2.2.4.1 Tissues

As an example, consider fixation using 4% paraformaldehyde in 10 mM PBS; pH 7.4.

↪ This may be considered the basic method.

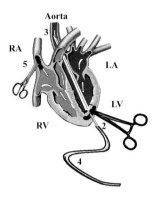

1 = **Opening of the thoracic cage and the pericardium**
2 = **Small incision in the left ventricle (LV)** near the tip of the heart
3 = **Introduction of the cannula** delicately up to the aorta
4 = **Clamping** of the cannula with the aid of artery clamps
5 = **Opening of the right auricle (RA)**

Figure 2.3 Method of perfusion.

1. Fixation by perfusion

↪ When the biological sample to be fixed is unstable, it is preferable to use intracardial perfusion (Figure 2.3).

a. Rinse with buffer.	**A few seconds**	↪ Allow ≈ 50 mL of preheated buffer per animal to pass through.
b. Administer the fixative to the anesthetized animal, at a slow rate, after rinsing.	**30 min**	↪ Allow 500 mL of fixative to perfuse each rat. The efficiency of perfusion can be checked by the stiffness of the animal's neck.
c. After perfusion, rapidly remove the organ and immerse in the same fixative.	**1 to 2 h at 4°C**	↪ The time of fixation varies according to the size of the sample, which must not be > 5 mm thick.
↪ For each fixative, there exists an optimal action time. There is no overfixation with aldehyde fixatives.
↪ A temperature of 4°C makes it possible to inhibit enzymatic activity, but fixation can also be carried out at room temperature.
d. Rinse with buffer	**2 × 15 min**	

2. Fixation by immersion

↪ See Figure 2.4.

a. Immediately suspend the biological sample in the fixative buffer. **8 to 24 h at 4°C**

↪ The tissue must be withdrawn as soon as possible after sacrificing the animal.
↪ Prepare a bottle for each type of sample. A cardboard triangle (carrying sample details, in pencil) should be placed in the buffer. This will accompany the sample throughout the handling procedures.

Tissue Preparation

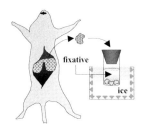

⇝ To identify the cassettes for embedding in paraffin wax, a special felt-tip pen may be used, or, if in doubt, a pencil.

⇝ The sample must fall to the bottom of the tube. If this does not happen, it is necessary to wait several minutes to allow air bubbles, which prevent the penetration of the fixative into the tissue, to be eliminated (e.g., with vegetable tissue, lung tissue).

⇝ It is possible to cut the sample a few minutes after hardening in the fixative.

Figure 2.4 Fixation by immersion.

b. Rinse with buffer	2 × 15 min at 4°C

⇝ It may be necessary to extend the fixation time for large samples.

⇝ A rinse time comparable to the fixation time is generally recommended, but this can damage the tissue. It is preferable to use numerous rinses, to eliminate the fixative completely.

Following steps — The fixed tissue can be:
- Frozen
- Cryoprotected and frozen
- Embedded in paraffin wax

⇝ *See* Section 2.3.4
⇝ *See* Section 2.3.3
⇝ *See* Section 2.4.3

2.2.4.2 Cells in suspension

1. Fixation

a. Rinse the cells using serum-free medium

b. After centrifugation of the cell suspension, remove the supernatant and replace immediately with fixative buffer.	10 min at 4°C

2. Rinsing

a. Remove the fixative after centrifugation, and immediately replace it with buffer.	3 × 5 min at 4°C

b. Keep the pellet in this buffer, and treat it like a tissue section.

Following steps — The fixed cellular pellet can be:
- Frozen
- Cryoprotected and frozen
- Embedded in paraffin wax

⇝ Cell cultures in monolayer after enzymatic treatment can be considered cells in suspension.

⇝ The presence of serum can create artifacts due to endogenous proteins.

⇝ It is sometimes necessary to preheat the fixation solution if the cells are in culture at 37°C, to preserve cellular morphology.

⇝ Fixation time is short, because the samples are only a few microns thick.

⇝ If necessary, centrifuge between rinses.

⇝ In certain cases, a coating in agarose at 1% facilitates the handling of the pellet.

⇝ *See* Section 2.3.4
⇝ *See* Section 2.3.3
⇝ *See* Section 2.4.3

2.2.4.3 Cells in culture/smears

1. Detachment of cells
a. An enzymatic or chemical treatment is necessary to detach the cells that are sticking to the plastic support, using one of the following washes:
- 0.25% trypsin in PBS **5 to 9 min at 37°C**

↪ In this case the number of cells is too low to obtain a pellet.

↪ Can lead to a morphological change

↪ With no bivalent ions

- 0.25% trypsin, 0.05% EDTA, in PBS **5 to 9 min at 37°C**
- EDTA in PBS

b. Rinse with buffer **3 × 5 min at 4°C**

↪ Possible to enhance its action by increasing the time
↪ Diluted 5:1000 ($^w/_v$)
↪ Centrifugation between rinses

2. Depositing the cells
— There are several ways of depositing the cellular suspension:
a. It can be **spread** directly onto pretreated slides.
b. It can be **projected** by cytocentrifugation onto pretreated slides.

↪ Smear.

↪ The technique of cytocentrifugation facilitates the gathering of cells from samples of all types (serum, urine, blood, pleural ascitic fluid, etc.).

- Cellular suspension **100 µL**

↪ The concentration of a cellular suspension is $\approx 1.5 \times 10^6$ cells / mL. Beyond that, the cells will be superimposed on one another, and will be removed.

- Centrifugation **1500 rpm for 10 min at RT**

↪ Centrifuge to check the homogeneity of the deposit. (RT = room + temperature)

- Drying **5 min at RT**

↪ Dry to fix the cells rapidly to preserve morphology.

3. Fixation
— The cells are immediately fixed after being deposited on slides:
- Fixative buffer **10 min at 4°C**

↪ The fixation technique is very rapid (the cells are spread in single layers).
↪ Cells cultured in single layers on histological slides should be placed directly in the fixative.

- Buffer **3 × 5 min**

Following steps — The fixed cells can be:
- Pretreated immediately before hybridization
- Stored in 70% alcohol until the hybridization step.
- Dehydrated, dried, and stored in airtight containers with a drying agent (e.g., silica gel) at –20°C

↪ *See* Chapter 3
↪ Complete elimination of unfixed aldehyde molecules
↪ Dehydration (70%, 95%, 100% alcohol) followed by drying essential before storage (drying is important for complete inhibition of the action of cellular RNases during storage)

Tissue Preparation

2.3 FROZEN TISSUE

Freezing is the change of cellular and / or tissue water from the liquid state to the solid state. This transformation leads to a physical "fixation" of the tissue, and allows sections to be cut.

2.3.1 Summary of the Different Steps

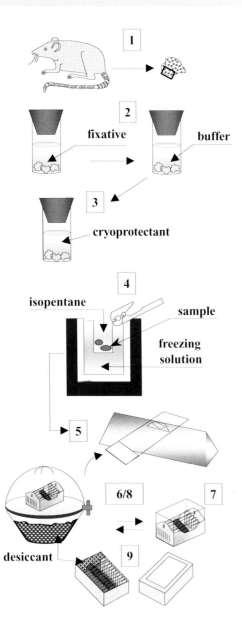

1 = Dissection (with or without fixation by perfusion)

2 = Fixation by immersion — Rinsing

3 = Cryoprotection (not always necessary) — 30 to 50% saccharose.

4 = Freezing (–160 °C) — storage (liquid nitrogen or freezer).

Never store in the cryomicrotome (thawing will take place during the night).

5 = Frozen sections — Freezing microtome or cryomicrotome (sections placed on pre-treated slides).
6 = Drying — sections may or may not be fixed.
7 = Dehydration — (alcohol baths of increasing strength).
8 = Drying

9 = Storage at –20 or –80°C (airtight containers with a drying agent).

Figure 2.5 Preparation of frozen tissue.

2.3.2 Equipment / Reagents / Solutions

2.3.2.1 Equipment
- Cryomicrotome

⇨ **Clean** (any contamination by other tissue can lead to problems caused by RNases)

- Freezer (– 20 or – 80°C)
- Quick freezing bath

⇨ For storing the sections
⇨ Enables freezing in isopentane (– 160°C) to take place without the formation of ice crystals

- Vacuum jar

⇨ For drying the sections

2.3.2.2 Small equipment

⇨ **Handle with gloves, and use equipment that has been sterilized / cleaned with ethanol.**

- Airtight containers (+ black tape)

⇨ For storing the slides; the container is hermetically sealed with black tape

- Dewar flask
- Glass staining trough
- Knife, disposable blades, brush, anti-roll device
- Metallic tweezers
- Pyrex beakers
- Slide frames

⇨ For freezing the tissue in liquid nitrogen
⇨ Sterilized (*see* Appendix A2)
⇨ Thoroughly cleaned in ethanol

⇨ Sterilized
⇨ For the isopentane
⇨ Slides placed in a vertical position to facilitate drying of the sections

2.3.2.3 Pretreated equipment
- Glass slides pretreated with silane

⇨ Essential treatment for the adhesion of the sections to the slide (*see* Appendix A3)

2.3.2.4 Reagents

⇨ **Molecular biology reagents are in general recommended, because they do not contain RNases. Reagents must be handled with gloves.**

- 95%, 100% alcohol
- Coating in tissue OCT-compound
- Desiccant

⇨ Neither too liquid nor too pasty
⇨ For example, silica gel

2.3.2.5 Solutions
- Formol
- 4% paraformaldehyde
- Phosphate buffer
- 30 to 50% saccharose

⇨ *See* Appendix B4.3
⇨ *See* Appendix B4.4.2
⇨ *See* Appendix B3.3
⇨ *See* Appendix B2.19

2.3.3 Cryoprotection

2.3.3.1 Principle
The cryoprotection step takes place between fixation and freezing. It limits the formation of ice crystals by replacing water with a solution where freezing will occur into an amorphous form (with the minimum amount of crystal formation).

⤳ **Not an essential step,** depending on the tissue

⤳ The more effective the cryoprotectant, the lower will be the freezing temperature, and the more difficult the section will be to cut, because the tissue will remain soft even at the temperature of the cryomicrotome.

2.3.3.2 Cryoprotectants
- 30 to 50% saccharose
- 5 to 10% glycerol

- 5 to 15% dimethylsulfoxide (DMSO)

- 15 to 50% polyvinylpyrolidone

⤳ Chemically the most neutral
⤳ Highly penetrating
⤳ A very good cryoprotectant, but cutting difficult because the sections remain soft
⤳ A good cryoprotectant, but an organic solvent which can alter membranes
⤳ A polymer that causes the transport of water out of the cell by hyperosmolarity

2.3.3.3 Method
The fixed and rinsed samples are placed in the cryoprotectant.
- Cryoprotectant   **2 to 24 h at 4°C**

⤳ The larger the sample, the more important is this step.
⤳ The tissue sample must fall to the bottom of the container. In this case, and only in this case, impregnation is total. This process can also, if necessary, be carried out in a partial vacuum.

2.3.4 Freezing

2.3.4.1 Principle
Cooling slows down all biological processes, and causes tissue hardening.
Homogeneous freezing constitutes one of the best methods of preserving nucleic acids *in situ*, and of preserving cellular morphology.

⤳ The hardness of the sections is an important parameter.
⤳ The size of the prepared tissue and the choice of the freezing solution are fundamental.

2.3.4.2 Freezing agents
- Liquid nitrogen (– 196°C)

⤳ **Danger, risk of burns**
⤳ Use adequate containers, i.e., not airtight **(risk of explosion).**

- Dry ice (– 78°C)
- Isopentane (2-methylbutane) cooled in liquid nitrogen (– 160°C)

- Propane

⇝ Not a good freezing solution. The layer of gas resulting from the vaporization of the liquid nitrogen causes an insulating zone between the freezing solution and the tissue.
⇝ Dry ice is a contact freezing agent
⇝ Below this temperature the isopentane becomes pasty, then solid. In this case, allow it to warm up before freezing the tissue.
⇝ The temperature remains constant for the duration of freezing.
⇝ Propane is an excellent freezing solution. **Danger, use with care.**

2.3.4.3 Method

Preparation of the samples — There are several ways to freeze samples:
- Direct freezing
- Freezing after fixation

- Freezing after fixation and cryoprotection

- Freezing after fixation and coating

⇝ Excess liquid on the surface of the tissue can form a shell of ice, which may lead to breaks. Dryness can cause morphological disruption.
⇝ The partial replacement of tissue water by a cryoprotectant changes the crystallization properties of the tissue fluid, which results in the formation of only very small crystals, or none at all.
⇝ Coating limits the risk of breakage, and gives to the biological sample the strength necessary for sectioning.
⇝ It is possible to place the prepared tissue and its OCT compound on the cryomicrotome support to facilitate freezing.

Freezing methods — There are several possible methods. The simplest is to place the sample manually:

1. In a liquid such as isopentane cooled by liquid nitrogen —freezing **1 min at – 160°C**
2. In nitrogen vapor, at the surface of the liquid nitrogen, before immersion in the liquid itself
3. Directly in liquid nitrogen

⇝ At –160°C the risks of breakage are minimal, and the thermal exchanges are excellent (the tissue is immersed in the freezing solution).
⇝ This may be advisable when freezing large samples, or when using a cryomicrotome.
⇝ There is a risk of breakage with large samples (formation of ice crystals in the case of muscle tissue).

Tissue Preparation

4. On dry ice

⇝ It is possible to place the sample on a support, or to wrap it in aluminum foil. In this case, allow it to float on the surface of the liquid nitrogen.

⇝ This enables several samples to be frozen at the same time. The area in contact with the dry ice will freeze best.

2.3.4.4 Storage

Frozen samples are stored until use, either:
- At – 196°C (liquid nitrogen)

- At – 20 or – 80°C

⇝ The lower the storage temperature, the longer the duration of preservation in good condition.

⇝ At – 80°C, and especially at – 20 °C, it is advisable to coat the tissue with an OCT compound, or to use another coating method. However, the preservation of these structures will always be of shorter duration. **It is preferable to fix this type of tissue before freezing.**

2.3.5 Preparation of Frozen Sections

2.3.5.1 Equipment

A cryomicrotome (Figure 2.6) is a microtome placed in a chamber whose temperature can be set at – 35°C so that the block and the knife are always at the optimal temperature for cutting. Cutting can be either manual or motorized.

⇝ The preparation of frozen sections can be carried out equally well whether or not the tissue has been fixed beforehand.

⇝ A cryomicrotome can be used to cut sections 3 to 50 µm thick.
1 = **Storage of sample supports** (mechanism for rapid freezing).
2 = **Sample support arm.**
3 = **Knife.**
4 = **Section.**
5 = **Collection tray** (for debris from sections).
6 = **Anti-roll device.**
7 = **Knife clamp.**

Figure 2.6 Cryomicrotome.

2.3.5.2 Protocol
1. *Mounting the block on the cryomicrotome support*

⇝ **The sample must never thaw out.** It is possible, if the sample is large, to mount it in liquid nitrogen vapor or on dry ice.

a. Place the frozen tissue directly in the refrigerated chamber of the cryomicrotome ($\approx -20°C$)

⇝ The choice of temperature is determined by the type of biological tissue:
- $\approx -12°C$ to $-25°C$
- Down to $-30°C$ for fatty tissue and bone (when it is necessary to use a tungsten blade).

b. Put the coating medium on the plate at room temperature, avoiding bubbles. It must perfectly fit the grooves in the support.

⇝ The concentric rings on the surface of the removable plate ensure perfect adhesion.

c. Put the support on the coldest part of the wall of the cryomicrotome.

⇝ This assures rapid freezing (see Figure 2.6).

d. Put the frozen sample on the support cooled to $\approx -20°C$, and coat it completely.

⇝ Deposit the sample as soon as the surrounding coating medium whitens.

e. Wait for 15 min before cutting the sections.

⇝ This allows the temperature of the sample to reach that of the cryomicrotome.

2. **Cutting frozen sections** — Several parameters determine the successful cutting of sections:
   - The cutting angle, starting at 6°
   - Static electricity

   - The cutting motion

   - The size of the block

   - The hardness of the sample
   - The sharpness of the knife
   - The temperature of the sample, the knife, and the chamber

⇝ **Gloves should be worn.**

⇝ Worked out by experimentation
⇝ Often a false problem, due to an ill-adjusted or dirty anti-roll device
⇝ The speed of the descent of the sample onto the edge of the knife
⇝ The smaller the block, the easier the sample will be to cut
⇝ The temperature of the tissue
⇝ Disposable blades are an alternative
⇝ These temperatures must be homogeneous. The lower they are, the thinner the sections can be, but in that case the tissue is more delicate.

The preparation of the block for obtaining sections comprises the following steps:

a. Make a cut in the shape of a trapezium, where the two parallel sides line up with the edge of the knife.
b. Adjust the knife.

⇝ If the sample is small, orient it so that the longer side is at the bottom (Figure 2.7).

⇝ Knife should be sterile, or at least cleaned in alcohol.

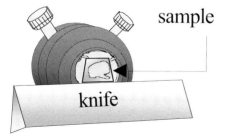

Figure 2.7 Cutting sections.

Tissue Preparation

c. Adjust the height of the sample by reference to the knife.

↪ **The block must not quite touch the cutting edge of the sample.**
↪ **To obtain good sections, the block and the knife must be parallel:**
• between the edge of the knife and the lower edge of the block;
• between the plane of cutting and the axis of the knife.

d. Attain the target area by cutting sections 25 μm thick.

↪ It is preferable to perform the operations behind the porthole; opening the refrigerated chamber will lead to condensation.

e. Obtain sections of a uniform thickness of 10 μm.

↪ Thinner sections will cause a reduction in the hybridization signal.
↪ If there is a problem of rolling up (too large a difference in temperature between the knife and the block), the temperature of the chamber must be changed.

f. Whatever the thickness of the sections, it is essential that they be well spread out.

↪ Set the anti-roll device against the blade to favor the spreading of the sections. The presence of folds in the tissue will lead to background noise at the time of hybridization.

g. At the time of cutting, the edge of the knife can be cleaned with a fine brush, or with a soft cloth soaked in ethanol.

↪ Never use a brush on the cutting edge of the knife.
↪ Wait until the ethanol has evaporated before cutting again.

3. Depositing the sections

The sections are deposited onto pretreated glass slides by simple contact with the edge of the knife. The difference in temperature between the sections and the slides allows the sections to adhere to the slides.

• Sections   1 or 2 / slide

↪ *Gloves should be worn.*
↪ For pretreated slides (*see* Appendix A3).
↪ Better spreading of the slides can be obtained by rubbing the underside of the slide with a finger immediately after cutting.
↪ Deposit the sections on the lower third of the slide.

4. Drying

Drying is essential before storing the slides

a. Without dehydration
• At room temperature   1 to 4 h
• In a vacuum jar   1 h

b. With dehydration

• 70% alcohol   3 min

↪ *Gloves should be worn.*
↪ The purpose of the drying step is to ensure adequate adhesion of the sections to the slide.
↪ RNases and DNases only function in the presence of water. Storage in the absence of water is the best way of inhibiting RNases.

↪ The sections must be subjected to total dehydration for **storage** of the slides.
↪ It is possible to store the slides in 70% ethanol until the pretreatment step.

• 95% alcohol	**3 min**
• 100% alcohol	**2 × 3 min**
• In a vacuum jar	**1 h**

5. Storage

a. Place the dried slides in airtight boxes containing a drying agent (e.g., silica gel).

↪ This form of storage can preserve the sections for several months, or indeed years.

↪ **Note:** Before using the sections, it is essential to leave them for **at least 2 h** at room temperature prior to opening the boxes (due to the risk of condensation, and thus rehydration of the sections).

b. Store   **RT**

↪ It is possible to store dehydrated sections at room temperature, which will not affect the duration of preservation.

– 20 or – 80°C

↪ These are the most frequently used temperatures.

Following steps — Preparations can be:
- Pretreated (chemically, or using enzymes)
- Prehybridized
- Hybridized

↪ *See* Chapter 3.
↪ *See* Chapter 3.
↪ *See* Chapter 4.

2.3.6 Advantages / Disadvantages

❏ *Advantages*
- Sensitivity
- Rapid obtainability of sections

↪ Optimum preservation of nucleic acids
↪ Absence of an embedding step

❏ *Disadvantages*
- Morphology
- Difficulty of storage

↪ Inferior to that of paraffin sections
↪ Necessitates a means of storage, and the duration is limited by comparison with paraffin blocks

- Preparation of sections

2.4 TISSUE EMBEDDED IN PARAFFIN WAX

Embedding in paraffin wax changes semiliquid biological material into a homogeneous solid that is sufficiently hard to enable sections to be cut, while also preserving tissue structures.

Tissue Preparation

2.4.1 Summary of the Different Steps

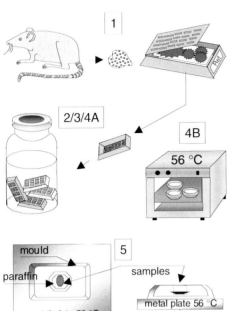

1 = Dissection (with or without fixation by perfusion) — Place in a cassette.

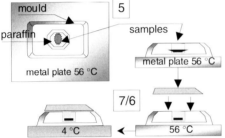

2 = Fixation by immersion — Rinsing.
3 = Dehydration (baths of alcohol of increasing concentrations).
4 = "Clearing" impregnation:
4A: solvent for paraffin
4B: liquid paraffin at 56°C

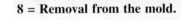

5 = Embedding (56°C) (embedding mold).

6 = Closure of the cassette.
7 = Cooling of the block.

8 = Removal from the mold.

9 = Cutting the sections — Microtome (5 to 10 µm).

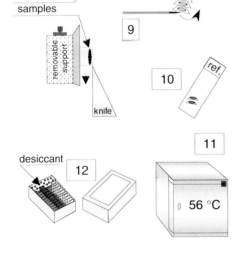

10 = Depositing the sections — on pre-treated slides.

11 = Drying (56°C).

12 = Storage (4°C).

Figure 2.8 Embedding in paraffin wax.

2.4.2 Equipment/Reagents/Solutions

2.4.2.1 Equipment

- Embedding system
- Hotplate
- Microtome for paraffin sections
- Oven
- Water bath

↪ Or, for paraffin wax, an oven
↪ To spread the sections
↪ Of the Minot type
↪ To dry the sections
↪ To recover the sections (the water must be distilled, sterilized, and regularly changed for each new set of sections)

2.4.2.2 Small equipment

↪ Handle with gloves, and use equipment that has been sterilized / cleaned with ethanol.

- Cassettes
- Embedding mould
- Glass staining trays

- Leuckart bars
- Needles, slide clip, brush
- Reusable or disposable knives

↪ To identify the tissue and move it
↪ Stainless steel or plastic
↪ Glass apparatus is autoclaved or sterilized (*see* Appendix A2)
↪ In the case of large samples
↪ For recovering the sections
↪ Hard metal or single-use knives

2.4.2.3 Pretreated equipment

- Glass slides pretreated with silane

↪ Essential for the adhesion of the sections to the slide (*see* Appendix A3)

2.4.2.4 Reagents

- Paraffin wax

↪ A mixture of aliphatic hydrocarbons that do not mix with water or alcohols but can be dissolved in xylene or a similar solvent

2.4.2.5 Solutions

- 70%, 95%, 100% alcohol
- Xylene or a similar solvent

2.4.3 Embedding in Paraffin Wax

2.4.3.1 Aim
The fixed biological material (tissue or cellular) is rinsed and successively subjected to:
- Dehydration
- Impregnation
- Embedding

1. **Dehydration** — In baths of ethanol at increasing concentrations.
2. **Impregnation** — By a solvent of the embedding medium.
3. **Embedding** — The most frequently used medium is paraffin.

↪ The success of embedding depends on the perfect dehydration of the samples, and the elimination of all traces of alcohol before impregnation.
↪ The ethanol replaces the water, as the two are totally miscible with each other.
↪ Paraffin is not miscible with ethanol, which must therefore be replaced by an intermediate liquid such as xylene or a similar solvent.
↪ There are other similar media.

2.4.3.2 Protocol

• NaCl 150 mM	3 min	
1. Dehydration	RT	
• 70% alcohol	1 × 30 min	
• 95% alcohol	1 × 30 min	
• 100% alcohol	2 × 1 h	
2. Impregnation		
• Xylene, or a substitute	2 × 15 min	
• Liquid paraffin	56 to 58°C	
– Tissue < 3 mm	2 × 90 min	
– Tissue > 3 mm	**Overnight**	
3. Embedding		
• Place in the cassette	56 to 58°C	

↪ **Optional step**
↪ This wash is necessary when using a phosphate buffer, to avoid a precipitate.
↪ Use ethanol baths of increasing concentrations.
↪ The duration of these steps must be increased in proportion to the size of the tissue.

↪ **Essential step**. The second bath must be fresh.
↪ The solvents used are toxic.
↪ The duration of impregnation varies according to the choice of intermediary solvent (the more volatile it is, the shorter will be the impregnation time). Impregnate the samples until they become translucent.
↪ In the case of reuse, hot filtration is essential.

↪ Each embedded sample must be referenced. Use a system that is indelible to solvents.
↪ Fill the cavity of the embedding mold (placed on a metal plate at 56°C) with liquid paraffin; position the sample at the bottom of the cavity and cover it. The cassette acts as a support to fasten to the arm of the microtome. The sample must be oriented. The bottom of the mold acts as the surface of the section.

2.4 Tissue Embedded in Paraffin Wax

- Use of Leuckart bars   **56 to 58°C (on a hotplate)**

⇝ This system allows blocks of varying dimensions to be obtained, using a set of bars (Figure 2.9).
⇝ The main advantage of this method is the embedding of bulky samples (thickness 5 mm).

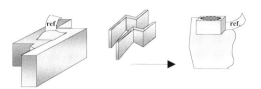

Figure 2.9 Leuckart bars.

4. Cooling the paraffin
5. Storage   RT

⇝ A freezer or ice box can be used.
⇝ A dry place is needed for long-term storage.

2.4.4 Preparation of Paraffin Sections

2.4.4.1 Equipment

All microtomes for paraffin sections (Figure 2.10) operate on the Minot principle, which comprises three essential features:
- Support for the object
- Support for the knife
- A movable mechanical stage graded in micrometers

The thickness of the sections varies from 5 to 10 μm

⇝ Their use requires little experience.

⇝ The stage is notched at each micron.
⇝ It is easier to obtain large sets of sections with paraffin-embedded than with frozen material.

1 = **Magnifying glass.**

2 = **Sample support arm.**

3 = **Knife clamp.**

4 = **Knife.**

5 = **Sections.**

6 = **Movable mechanical stage.**

Figure 2.10 Microtome for paraffin sections.

2.4.4.2 Protocol

1. Obtaining paraffin sections

Several parameters are involved in obtaining the section:

- The positioning of the knife

↪ Angle should be between 10 and 15° (angle between the knife and the surface of the section; Figure 2.11)

1 = Paraffin wax block.
2 = Angle of inclination.
3 = Knife.

Figure 2.11 Angle of inclination of the knife.

- The cutting motion

↪ The speed of descent of the sample onto the edge of the knife is determined empirically, as it is dependent on the block.

- The size of the block

↪ The smaller the surface area of the sections, the easier it is to obtain fine sections, but this is not essential for *in situ* hybridization. The quantity of target nucleic acid detected is generally proportional to the thickness of the section.

- The knife (or disposable blade)

↪ It is cleaned in alcohol.
↪ Any cutting artifacts caused by the knife (e.g., ridges or holes) will lead to nonspecific signals.

- Static electricity

↪ This is a phenomenon that can be eliminated by the use of antistatic accessories.

The preparation of the block for making sections comprises the following steps:

a. Use a trapezium block whose two parallel sides are parallel with the cutting edge. The longer side is at the bottom of the trapezium.

↪ The two other faces enable the sections to be distinguished from each other, and make their separation easier. The surface of the section must be kept as small as possible by having 2 to 3 mm paraffin around the sample (Figure 2.12).

Figure 2.12 Sizing the paraffin wax block.

b. Adjust the approach of the block in relation to the knife.

↪ Position the block parallel to the cutting edge to maximize the surface area of the section.

c. Reach the sample by means of a rapid advance of the stage, then cut sections 10 μm thick until a whole-sample section is obtained.

↪ Set the micrometer at 10 μm and advance the stage rapidly until the first shaving is obtained (trimming). If necessary, reorient to obtain a whole-sample section.

d. Set the required thickness.

➥ Obtain sections of regular thickness between 5 and 10 μm.

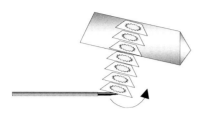

e. At the time of cutting, the edge of the knife can be cleaned with a xylene wet paper.

f. Prepare a strip of five to eight sections.

g. Check the quality of the sections.

• Spread the sections.   40 to 45°C

2. *Depositing the sections*
                              1 to 2 sections / slide

3. *Drying*
                              Overnight
                              56°C oven

4. *Storage* — place the dried slides in airtight boxes containing a desiccant (e.g., silica gel)   4°C

2.4.5 Dewaxing the Sections

2.4.5.1 Aim
Dewaxing consists of the elimination of the embedding medium.
It makes rehydration of the tissue possible.

This step is carried out **just before** pretreatment.

Figure 2.13 Obtaining a strip of paraffin sections.

➥ The cutting edge is cleaned with xylene. Before cutting again, wait until the xylene has completely evaporated. Never use a brush on the cutting edge of the knife.
➥ With a scalpel, support the free end of the strip and divide it into sections by sliding a brush under the last raised section (Figure 2.13).
➥ To judge the quality of sections during preparation, remove one section from the set and examine it under a light microscope.
➥ On the surface of a water bath, sterile water at 40 to 45°C. The temperature of the water must be high enough to enable the spreading of the sections to take place, but not high enough for the paraffin wax to melt.
➥ *Gloves should be worn.*
➥ Use pretreated slides (*see* Appendix A3)
➥ Deposit one or more sections on the lower third of the slide (slide + 50% alcohol to favor the smoothing out of the sections).
➥ Place the slide and sections on a hot plate at 45°C.
➥ *Gloves should be worn*
➥ The purpose of the drying step is to ensure sufficient adhesion of the sections to the slide (minimum time 2 h at 56°C).
➥ This is not necessary but preserves the surface of the sections against RNase effects.

➥ **Essential step**, carried out just before pretreatment (*see* Section 3.4.5).

➥ Hydrophobic medium

➥ **Essential** before using permeabilizing treatments.
➥ **Storage not possible after dewaxing**

2.4.5.2 Method

1. Dewaxing
   - Xylene or a substitute   **2 × 5 min**   ↪ Baths must be regularly changed.
2. Rehydration   **RT**
   - 100% alcohol   **3 min**
   - 95% alcohol   **3 min**
   - 70% alcohol   **3 min**
   - NaCl 150 mm   **3 min**   ↪ Na^+ ions keep the nucleic acids *in situ*. Avoids the need for washes in water.

Following steps — Preparations can be:
- Pretreated immediately before hybridization   ↪ *See* Chapter 3.
- Hybridized   ↪ *See* Chapter 4.

2.4.6 Advantages / Disadvantages

❏ *Advantages*
- Preservation of morphology   ↪ All morphological structures are maintained after embedding.
- Ease of obtaining sections
- Easy storage   ↪ It is preferable to store blocks rather than sections.

❏ *Disadvantage*
- Weak sensitivity   ↪ This technique does not allow low levels of mRNA to be detected in tissue.

Chapter 3

Pretreatments

Contents

3.1	Principle	93
3.2	Summary of the Different Steps	94
3.3	Equipment/Reagents/Solutions	95
	3.3.1 Equipment	95
	3.3.2 Reagents	95
	3.3.3 Solutions	96
3.4	Stabilization of Structures	97
	3.4.1 Aim	97
	3.4.2 Protocol	97
3.5	Permeabilization	98
	3.5.1 Aim	98
	3.5.2 Agents	98
	3.5.3 Protocols	98
3.6	Deproteinization	99
	3.6.1 Aim	99
	3.6.2 Enzymatic Treatment	99
	3.6.2.1 Proteinase K	99
	3.6.2.1.1 Utilization	99
	3.6.2.1.2 Protocol	100
	3.6.2.2 Other Enzymes	100
	3.6.2.2.1 Utilization	100
	3.6.2.2.2 Protocols	101
	3.6.3 Chemical Treatment	102
3.7	Acetylation	102
	3.7.1 Aim	102
	3.7.2 Protocol	102
3.8	Dehydration	102
	3.8.1 Aim	102
	3.8.2 Protocol	103
3.9	Prehybridization	103
	3.9.1 Aim	103
	3.9.2 Protocol	103
3.10	Denaturation of the Target Nucleic Acid	104
	3.10.1 Aim	104
	3.10.2 Protocols	105
	3.10.2.1 Physical Treatment	105
	3.10.2.2 Chemical and Physical Treatment	106

3.1 PRINCIPLE

The aim of pretreatments is to make a **target nucleic acid** accessible to a **probe**, that is to say, to complementary sequences of nucleic acids. The cell components, in close association with the nucleic acids, are essentially proteins and nucleic acids.

Pretreatments destroy or modify these proteins or their functional groups, thus avoiding a possible attachment of the probe to the proteins in the preparation.

They can also denature the target nucleic acid and the probe if necessary.

Pretreatments include carrying out controls, along with improvements in the signal/background ratio and the hybridization signal.

⇝ These treatments are optional, and depend on the origin of the biological material and fixative used, but also on the necessity of increasing the signal.

⇝ These are deproteinization and permeabilization steps.
⇝ These are acetylation and prehybridization steps.
⇝ This is the denaturation step.

⇝ These improvements are mainly the result of increasing the possibilities of hybridization, and of limiting the possibilities of nonspecific hybridizations.

3.2 SUMMARY OF THE DIFFERENT STEPS

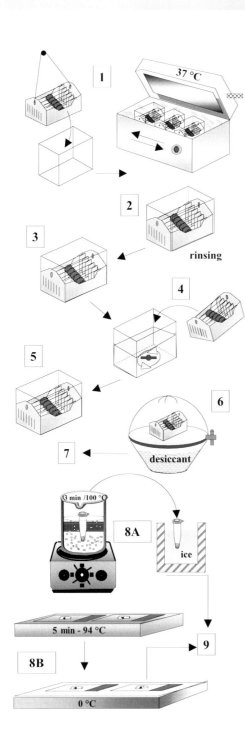

Storage at – 20°C
Leave for 2 h at room temperature before opening the box (to avoid all risk of condensation).

1 = Rehydration
- Fixation if necessary
- For paraffin-embedded sections, permeabilization
- Rinsing

2 = Deproteinization — Rinsing

3 = Postfixation — Rinsing

4 = Acetylation (optional)

5 = Dehydration — Series of alcohols of increasing concentrations

6 = Drying

7 = Prehybridization

Dehydration (optional)

8 = Denaturation

8A: **The probe**
8B: **The target nucleic acid**

9 = *In situ* **hybridization**

Figure 3.1 The pretreatment steps.

3.3 EQUIPMENT/REAGENTS/SOLUTIONS

3.3.1 Equipment

- Centrifuge
- Heating block with thermostat ≈ 92 to 100°C

- Micropipettes
- Microwave 245 mHz, ≈ 750 W
- Shaking device

- Slide pincers

- Staining trays + slide-holders

- Vacuum jar or Speedvac

- Vortex
- Water bath, with shaking

↪ > 14,000 g
↪ The temperature is important, because it determines the denaturation time and the preservation of tissue morphology.

↪ The quality of the microwave is important.
↪ Horizontal shaking should be of low amplitude.
↪ Sterilized pincers should be wrapped in aluminum foil and stored at room temperature.
↪ Trays and holders should be pyrex (resistant to dry sterilization at 180°C). Equipment is to be used only for *in situ* hybridization or disposable.
↪ It must be large enough to take the slide-holders.

↪ 30 to 70°C.

3.3.2 Reagents

- Acetic anhydride
- Bipotassium phosphate
- Bovine serum albumin (BSA)
- Calcium chloride
- Dextran sulfate
- Digitonin
- Disodium phosphate
- DNA

- Ethylene diamine tetra-acetic acid
- Ficoll 400
- Formamide
- Glutaraldehyde
- Hydrochloric acid
- Monosodium phosphate
- NaOH
- Paraformaldehyde
- Pepsin
- Pronase
- Proteinase K
- RNA
- Saponin
- Silica gel

↪ **To be used only for *in situ* hybridization**
↪ AR quality
↪ Very high quality
↪ In powder
↪ Fraction V

↪ Exists in powder or solution

↪ In powder
↪ From salmon or herring sperm, in solution/sonicated
↪ Titriplex III (EDTA)
↪ Molecular biology quality
↪ Exists in deionized form
↪ Commercial solution, 25 to 70%

↪ In powder
↪ To be stored in dry conditions
↪ In powder
↪ In powder or solution
↪ In powder or solution
↪ Molecular biology quality
↪ Transfer RNA, in solution

↪ Desiccant, can be regenerated

- Sodium chloride
- Triethanolamine
- Tris-hydroxymethyl-aminomethane
- Trisodium citrate
- Triton X-100
- Xylene or a substitute

3.3.3 Solutions

- Alcohol 50%, 70%, 95%, and 100%
- Buffers
  — 100 mM triethanolamine buffer; pH 8.0

  — Phosphate buffer 10X; pH 7.4

  — Hybridization buffer
  — PBS 1X or 10X

  — TE (100 mM Tris–HCl; 10 mM EDTA) 10X, pH 7.6
  — TE/NaCl 10X (100 mM Tris–HCl; 10 mM EDTA, NaCl 5 M); pH 7.6
  — Tris–HCl / $CaCl_2$ (20 mM Tris, 2 mM $CaCl_2$); pH 7.6
  — Tris–HCl / glycine (50 mM Tris, 50 mM glycine); pH 7.4
- 2 mM calcium chloride

- 50% dextran sulfate
- DNA 10 mg / mL

- Deionized formamide

- Denhardt's solution 50X

- Digitonin 0.05%
- NaOH 10 N

- 40% paraformaldehyde

- 4% paraformaldehyde (PF) PBS
- Pepsin
- Poly A 10 mg / mL
- Polyethylene glycol (PEG) 10 mg / mL

↝ Anhydride

↝ In powder, stored in dry conditions
↝ Anhydride
↝ Stored in darkness
↝ Histological quality

↝ **The sterility of the solutions is crucial for the hybridization process** (*see* Appendix A2).

↝ *See* Appendix B4.2

↝ Storage at 4°C or room temperature (*see* Appendix B3.1)
↝ Storage at room temperature for 1 month after sterilization (*see* Appendix B3.3)
↝ Without the probe
↝ Storage at room temperature after sterilization (*see* Appendix B3.3.3)
↝ Storage at 4°C or room temperature (*see* Appendix B3.5.1)
↝ Storage at 4°C or room temperature (*see* Appendix B3.5.2)
↝ Storage at 4°C or room temperature (*see* Appendix B3.6.2)
↝ Solution 1X (*see* Appendix B3.6.4)

↝ Storage at room temperature after sterilization (*see* Appendix B2.5)
↝ Storage at – 20°C (*see* Appendix B2.22)
↝ Sonicated, storage at – 20°C; the freezing/thawing cycles can cause breaks (*see* Appendix B2.3).
↝ Storage at – 20°C in aliquots of 500 μL; if the reagent is not frozen at – 20°C, do not use it (*see* Appendix B2.13).
↝ Storage at – 20°C in aliquots; several thawings are possible without any modification of properties (*see* Appendix B2.9).
↝ Dilution in phosphate buffer
↝ Do not use if precipitated (*see* Appendix B2.21)
↝ Storage at room temperature for 1 month (*see* Appendix B4.4.1)
↝ Storage at – 20°C (*see* Appendix B4.4.2)
↝ Diluted in HCl 0.2 N; pH 5.0
↝ Equivalent of RNA (*see* Appendix B2.14)
↝ *See* Appendix B2.15

- Pronase 0.1 mg / mL Tris–HCl buffer
⇝ Store at −20°C in aliquots of 250 μL (*see* Appendix B2.16).

- Proteinase K 1 mg / mL sterile water
⇝ Store at −20°C in aliquots of 250 μL (*see* Appendix B2.17).

- RNA 10 mg / mL
⇝ Store at −20°C; the freezing/thawing cycles can cause breaks (*see* Appendix B2.4).

- RNase 100 μg / mL sterile water
⇝ Store at −20°C in aliquots of 40 μL.
⇝ To be activated, the solution must be boiled for 5 to 10 min (*see* Appendix B2.18).

- Saponin 0.1%
⇝ Dilute in sterile water.
- 150 mM sodium chloride
⇝ Store at room temperature after sterilization (*see* Appendix B2.8).

- Sterile water
⇝ *See* Appendix B1.1; use only at the time of opening the bottle, or use DEPC-treated water (*see* Appendix B1.2).

- SSC 20X; pH 7.0
⇝ Store at 4°C or room temperature (*see* Appendix B3.4).

- 0.1% Triton X-100
⇝ Dilute in sterile water (*see* Appendix B2.24).

3.4 STABILIZATION OF STRUCTURES

3.4.1 Aim

Fixation is intended to:
- Preserve morphology
- Stabilize the functional groups of the cell constituents
- Improve the adhesion of the sections to the slides
- Improve the efficacy of the hybridization

If the samples or sections have **not been fixed** (in the case of frozen tissue sections, cytocentrifuged cells, or smears), fixation is indispensable before hybridization.

⇝ *See* Chapter 2.
⇝ In the case of deproteinization treatment, it is preferable to carry out pre- and postfixation before and after the treatment.

3.4.2 Protocol

Dip the slides in the following baths:
1. Rehydration
   - 150 mM NaCl   **3 min**

   ⇝ A single rehydration step in the fixation buffer may be enough if the tissue has been dried in air.

   - 100 mM phosphate buffer   **3 min**
2. Fixation
   - 4% PF in phosphate buffer   **15 min at RT**

   ⇝ Other fixatives can be used (e.g., ethanol, alcohol–acetone, or formol).

Pretreatments

3. Rinsing — Phosphate buffer 2×5 min
4. Drying

⇝ After rinsing, dry the slides at room temperature only in the case of fixation with alcohol; then store at $-20°C$

The preparations can be:

- Deproteinized
- Prehybridized
- Hybridized
- Stored at $-20°C$ after dehydration and drying in hermetic boxes in the presence of a desiccant

⇝ *See* Section 3.6.
⇝ *See* Section 3.9
⇝ *See* Chapter 4.
⇝ These is no apparent reduction in the hybridization signal after more than 2 years if the opening of the storage boxes is carried out in the absence of humidity (*See* Appendix A1.1).

3.5 PERMEABILIZATION

3.5.1 Aim

Permeabilization a useful step in the case of very thick sections, and for antigenic probe revelation (penetration of the reagents in the immunohistological reaction).

3.5.2 Agents

- Detergents (SDS, Triton X-100, saponin)

- Digitonin

⇝ In the case of cells in culture, and paraffin-embedded sections, saponin can replace proteinase K, while also providing permeabilization and deproteinization.
⇝ In the case of paraffin-embedded sections, treatment with digitonin can be carried out before treatment with proteinase K, for better diffusion of the probe in the tissue.

3.5.3 Protocols

Dip the slides in one of the following baths:

⇝ At room temperature

- 0.05% digitonin **10 min**
- 0.1% Triton X-100 **5 to 15 min**
- 0.1% saponin **5 to 15 min**

After rinsing (PBS), the preparations can be deproteinized.

⇝ *See* Section 3.6.

3.6 DEPROTEINIZATION

3.6.1 Aim

The deproteinization treatments are intended to partially eliminate proteins and, more particularly, those associated with nucleic acids, to give the probes better access to the target complementary sequences. Three pretreatments are possible:

⇝ This treatment can be carried out when the sections are obtained; the sections can be stored in deproteinized form.

- Enzymatic
- Chemical
- Physical

⇝ With Proteinase K, pronase, pepsin, etc.
⇝ With HCl, NaOH
⇝ With detergents that cause partial deproteinization (*see* Section 3.5)

3.6.2 Enzymatic Treatment

3.6.2.1 Proteinase K

Proteinase K is generally considered to be a selective protease for proteins associated with nucleic acids.

⇝ This pretreatment remains a compromise between the intensity of the signal and the preservation of the appearance of the tissue.

3.6.2.1.1 UTILIZATION

The effect of proteinase K is modified by the following parameters:

⇝ This step is crucial to the success of the reaction.

1. *Concentration* — The enzyme concentration can vary according to the type of sample (cells, frozen sections, or paraffin-embedded sections), but also according to the nature of the tissue and the target:

  - Cytocentrifuged cells   1 µg / mL
  - Frozen sections   1 to 3 µg / mL
  - Paraffin-embedded sections   5 to 10 µg / mL

  ⇝ The single layer of cells must be treated with care.
  ⇝ Moderate digestion (1 µg/mL) gives good results. A higher concentration can alter tissue morphology.
  ⇝ Variations occur according to the fixation and thickness of the sections.

2. *Buffer* — Proteinase K (stock solution) is diluted in 20 mM Tris buffer; 2 mM $CaCl_2$; pH 7.5

  ⇝ This buffer contains calcium, which is the cofactor of proteinase K, but it is possible to use other buffers (e.g., TE *see* Appendix B2.12) to limit the action of the proteinase K.

Pretreatments

3. ***Temperature*** — The optimal temperature is **37°C**.
4. ***Duration***   **5 to 30 min**

↪ The enzyme remains active at lower temperatures.
↪ Duration should vary according to the specific activity of the enzyme and to the thickness of the section.

3.6.2.1.2 PROTOCOL

1. Rehydrate the preparations before the treatment (NaCl, phosphate buffer).
2. Place the Tris / $CaCl_2$ buffer in a water bath, and dip the slides when the buffer is at 37°C
3. Enzymatic treatment   **15 min at 37°C**
 — Proteinase K
Add the proteinase K at the last moment, and shake with the slide tray.
4. Stopping the reaction:
   - Tris / $CaCl_2$ **2 min**
   - Phosphate buffer   **5 min**

↪ This is necessary if the fixed sections were stored (– 20°C)
↪ The choice of buffer modifies the activity of proteinase K
↪ Use a water bath at 37°C, with shaking

↪ The previously cited parameters can modify these effects.

↪ Rinse rapidly.
↪ This change of buffer stops the action of the proteinase K, and eliminates the NH_2 groups from the Tris buffer, and also the calcium chloride.

5. Postfixation

   - 4% PF in phosphate buffer   **5 min**

↪ Restabilization of the tissue is **indispensable before hybridization.**
↪ This fixative gives the best results for the ratio of efficacy of hybridization to preservation of the morphological appearance.
↪ This is necessary to avoid the sections coming loose during the hybridization step.

6. Rinsing
   - Phosphate buffer   **3 × 5 min**
   - 150 mM NaCl   **3 min**

↪ This eliminates any trace of fixative.
↪ This is necessary to avoid precipitates in the alcohol.

Following steps — The preparations can be:
   - Acetylated
   - Dehydrated and dried

↪ If the probes are cRNA; *see* Section 3.7.
↪ *See* Section 3.8.

3.6.2.2 Other enzymes

It is possible to use other enzymes, such as pronase or pepsin.

↪ They are more difficult to use. Their activity depends on the origin of the enzyme, and even, sometimes, that of the batch.

3.6.2.2.1 UTILIZATION

1. ***Concentration***

↪ The concentration depends on the thickness of the sections, and on the fixation.

- *Pronase*
  - Cytocentrifuged cells **20 to 100 µg / mL**
  - Frozen sections **20 to 100 µg / mL**
  - Paraffin-embedded **1 to 4 mg / mL**
    sections
- *Pepsin*
  - Cytocentrifuged cells **5 to 10 µg / mL**
  - Frozen sections **5 to 10 µg / mL**
  - Paraffin-embedded **1 to 4 mg /mL**
    sections

2. *Buffer*
   - *Pronase* — TE buffer **pH 7.6**
     (Tris – HCl 10 mM,
     EDTA 1 mM)
   - *Pepsin/HCl* **pH 5.0**

3. *Temperature*
   - *Pronase* **RT or 37°C**

   - *Pepsin* **RT or 37°C**

4. *Duration*
   - *Pronase*
     - Cytocentrifuged cells **1 to 2 min**
     - Frozen sections **1 to 30 min**
     - Paraffin-embedded **1 to 30 min**
       sections
   - *Pepsin*
     - Cytocentrifuged cells **15 min**
     - Frozen sections **15 to 60 min**

     - Paraffin-embedded **15 to 60 min**
       sections

3.6.2.2.2 PROTOCOLS
1. *Pronase*
   a. Rinse in Tris buffer
   b. Incubate in the pronase solution
   c. Rinse in Tris – glycine **5 min**
      buffer
   d. Rinse in 150 mM NaCl **5 min**
2. *Pepsin*
   a. Rinse in Tris buffer.
   b. Treat the slides with pepsin diluted in HCl
      0.2 N; pH 5.0.
   c. Rinse in Tris buffer; **5 min**
      pH 8.0.

↪ Activity is variable from one batch to another.

↪ Pepsin is a pH-dependent enzyme.

↪ Pronase can be used with other buffers (e.g., 50 mM Tris–HCl).

↪ Pepsin is active at acid pH. Its activity can be reduced by raising the pH.

↪ The effect of this protease is highly dependent on the temperature, and on the batch of enzyme.
↪ **Use with care.**
↪ Its optimal temperature is 37°C, but it remains active at room temperature.

↪ This is an extremely delicate step.

↪ Duration is greater for frozen tissue.

↪ Time varies according to the temperature and pH.

↪ This solution is prepared at time of use.
↪ Stop the action of the pronase with the Tris – glycine buffer.
↪ The sections can then be dehydrated in alcohol.

↪ The concentration, the duration, and the temperature must be checked for each batch.
↪ Check the pH, which must be basic.

Following steps — The preparations can be:
- Acetylated
- Dehydrated and dried

↪ If the probes are cRNA, *see* Section 3.7.
↪ *See* Section 3.8.

3.6.3 Chemical Treatment

↪ This is an alternative treatment to enzymatic proteolysis, essentially used in the case of viral DNA detection.

1. Incubate the preparations in HCl 0.2 *N* solution. **10 min**
2. Rinse in distilled water. **2 to 5 min**
3. Let the slides drip-dry.
4. Dehydrate, dry.

↪ Very stable solution
↪ Room temperature

↪ *See* Section 3.8.

3.7 ACETYLATION

↪ This is very useful with cRNA probes.

3.7.1 Aim

The aim is to transform the reactive amine group (NH_3^+) of proteins into a substituted amine group ($-NH-CO-CH_3$, neutral) by reacting the sections with acetic anhydride ($CH_3-CO-CH_3$) in a triethanolamine buffer at pH 8.0.

↪ This reaction is effective **in reducing background**, although the reaction also acts on the nitrogenous bases of nucleic acids, and can cause a reduction in the hybridization signal.

3.7.2 Protocol

1. Dip the slides in a tray containing 200 mL triethanolamine buffer, and a magnetic rod for shaking
2. Shake, then add:

   - Acetic anhydride **1 mL**
3. Incubate **1 min**

   - Phosphate buffer **5 min**

   - 150 m*M* NaCl **5 min**
4. Dehydrate

↪ The acetylation of the preparations is carried out before the denaturation of the DNA.
↪ The shaking is crucial, given that the reaction is extremely rapid.
↪ Final concentration is 0.5%.
↪ The reaction takes place in less than 1 min with **constant shaking.**
↪ One bath is enough to eliminate all trace of triethanolamine.

↪ *See* Section 3.8.

3.8 DEHYDRATION

3.8.1 Aim

Dehydration eliminates the aqueous medium from the section to prevent any dilution of the hybridization buffer or the probe.

3.8.2 Protocol

1. Dehydrate the preparations in a series of alcohols of increasing strengths at room temperature:
   Alcohol
   70%, 95%, 100% **2 min / bath**
2. Dry
   - In a vacuum jar **> 30 min**

   - In air **> 60 min**

   Following steps: The preparations can be:
   - Prehybridized
   - Hybridized
   - Stored at −20°C, after dehydration and drying, in hermetic boxes in the presence of a desiccant.

↪ A bath of NaCl (150 m*M*) generally prevents all interference between the previous buffer and the alcohols.

↪ The enzyme (RNase, DNase) are inactive in the absence of water.
↪ 60 min is the minimum time.

↪ *See* Section 3.9.
↪ *See* Chapter 4.
↪ *See* Appendix A1.1.

3.9 PREHYBRIDIZATION

↪ Prehybridization is useful in the case of paraffin-embedded sections, and optional in that of frozen tissue.

3.9.1 Aim

The aim is saturation of nonspecific sites that can adsorb a probe labeled by macromolecules contained in the buffer (dextran sulfate, Denhardt's solution, RNA, DNA).

↪ This pretreatment step is not indispensable, but it is useful in reducing background noise.
↪ The prehybridization step is carried out before the hybridization step. It consists of incubating sections in a hybridization medium, without the probe.

❑ *Advantage*
- Reduction of nonspecific signals

❑ *Disadvantage*
- Reduction in sensitivity

↪ Risk of diluting the hybridization buffer and the probe during the hybridization process as such.

3.9.2 Protocol

1. After dehydration and drying, the sections are incubated with the hybridization medium, without the probe.
   - Hybridization buffer **≈ 100 µL**

↪ *See* Chapter 4. It is possible to increase the concentrations of DNA, RNA, and Denhardt's solution, in place of the formamide.

| • Incubation time | 1 to 2 h RT | ↬ This can be a waiting step (some hours). |

↬ This temperature does not modify non-specific bonds (weak influence of formamide).

2. The maximum amount of buffer is eliminated before placing the labeled probe and hybridizing.

Following step — The preparations may be hybridized.

↬ The sections must not be allowed to dry. They should be left moist to favor hybridization.

↬ *See* Chapter 4.

3.10 DENATURATION OF THE TARGET NUCLEIC ACID

↬ Only for double-stranded sequences

3.10.1 Aim

For some target nucleic acids, because they are double stranded, or have intrachain matchings, denaturation is necessary to obtain hybridization with the complementary single-stranded sequences contained in the hybridization solution (probe).

The simultaneous denaturation of the nucleic acid target and the probe is possible.

There are several methods for denaturing intracellular double-stranded structures:

- Physical (by heat)
- Chemical and physical (in the presence of formamide, and with heat)
- By alkaline treatment (NaOH)

1. ***Physical, by heat*** — The separation of the two strands of a double-stranded nucleic acid is generally carried out by thermal denaturation (Figure 3.2).

Pretreatment by heat can be used to improve the detection of cellular RNA. This step is carried out just before the *in situ* hybridization step.

↬ Although the double-stranded structure of DNA is obvious, it is always useful to check the structure of the target nucleic acid in order to find out if such matchings are possible or if they exist.

↬ In this case, the probe is placed on the sections and the combination is denatured.

↬ What is sought is the best compromise between the signal/noise ratio and the preservation of morphology.

↬ NaOH is only used in electron microscopy.

↬ The matching of two strands of DNA is an exergonic reaction. Further energy will be needed to separate the two strands.

↬ This technique favors the diffusion of the probe.

3.10 Denaturation of the Target Nucleic Acid

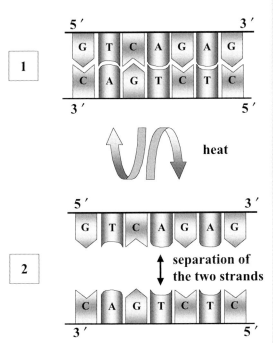

Figure 3.2 Principle of denaturation.

❑ *Advantages*
- Simple
- Rapid
- Reliable, with a reduction of background noise

❑ *Disadvantages*
- Risk of evaporation and drying
↝ An increase in background
- Alterations in morphology

2. **Chemical and physical** — The addition of formamide to the hybridization buffer makes it possible to lower the denaturation temperature.

↝ The addition of 1% formamide to the hybridization buffer reduces the denaturation temperature by 0.65°C.

❑ *Advantages*
- Preservation of morphology
- Limited risk of drying

❑ *Disadvantage*
- Time-consuming

3.10.2 Protocols

3.10.2.1 Physical treatment

1. Place the hybridization buffer containing the probe on the sections.
2. Place the slides on the **heating block.** The sections can be covered by a coverslip and sealed.

↝ The heating block must be maintained at the correct temperature.
↝ It is possible to denature the target and the probe simultaneously.

Pretreatments

The treatment can be carried out either at:

• 100°C	**3 min**	➢ There is risk of degradation of tissue structures.
• 94°C	**6 min**	➢ This is a good compromise between temperature and duration.
• 92°C	**10 min**	➢ Care must be taken in drying the tissue.

➢ **Alternative**: Heat the slides in a microwave oven (750 W) 3 × 3 min, with 3 min breaks.

Following step — The preparations are then hybridized.

➢ *See* Chapter 4.

3.10.2.2 Chemical and physical treatment

1. Make up a hybridization buffer without a probe containing 70% formamide. Place the hybridization buffer containing the probe on the sections.

   ➢ This buffer contains the SSC solution and the DNA.

2. Heat on the heating block:
   - 80°C **6 min**
   - 85°C **2 min**
3. Place the slides on ice.
4. Dehydrate in cold ethyl alcohol:
   - 95% **2 min**
   - 100% **2 × 2 min**
5. Dry in air.

Following step — The preparations are then hybridized.

➢ *See* Chapter 4.

Chapter 4

Hybridization

Contents

4.1	Principle of *In Situ* Hybridization.		111
4.2	Parameters		114
	4.2.1	Structure of Nucleic Acids	114
		4.2.1.1 Nucleoside.	114
		4.2.1.2 Nucleotide.	115
		4.2.1.3 DNA	116
		4.2.1.4 RNA	117
		4.2.1.5 Duplication	118
		4.2.1.6 Replication	118
		4.2.1.7 Transcription	119
	4.2.2	Complementarity of Bases	119
	4.2.3	Hybridization Temperature.	120
		4.2.3.1 Determination of Tm.	120
		4.2.3.2 Determination of the Hybridization Temperature.	121
	4.2.4	Na^+ Concentration	122
	4.2.5	Hybridization Buffer.	122
	4.2.6	Type of Probes	122
	4.2.7	Length of Probes.	122
	4.2.8	Type of Label Used.	123
	4.2.9	Homology of Sequence	123
4.3	Posthybridization Treatments.		123
	4.3.1	Purpose of Washing	123
	4.3.2	Stringency Conditions	124
		4.3.2.1 Determination of the Tw	124
		4.3.2.2 Ionic Strength.	125
		4.3.2.3 Washing Time	125
4.4	Summary of the Different Steps		126
4.5	Equipment/Reagents/Solutions.		127
	4.5.1	Equipment.	127
	4.5.2	Reagents	127
	4.5.3	Solutions.	128
4.6	Composition of the Hybridization Medium		129
4.7	Protocols.		130
	4.7.1	Protocol for cDNA Probe	130
		4.7.1.1 Hybridization Buffer.	130
		4.7.1.2 Reactive Medium	130
		4.7.1.3 Hybridization	131
		4.7.1.4 Washing.	132
	4.7.2	Protocol for Oligonucleotide Probe	132
		4.7.2.1 Hybridization Buffer.	132
		4.7.2.2 Reactive Medium	133
		4.7.2.3 Hybridization	133
		4.7.2.4 Washing.	134

	4.7.3	Protocol for cRNA Probe	134
		4.7.3.1 Hybridization Buffer	134
		4.7.3.2 Reactive Medium	135
		4.7.3.3 Hybridization	135
		4.7.3.4 Washing	135
4.8	Before Revelation		136
	4.8.1	Radioactive Label	137
		4.8.1.1 Dehydration	137
		4.8.1.2 Drying	137
	4.8.2	Antigenic Label	137

4.1 PRINCIPLE OF *IN SITU* HYBRIDIZATION

Hybridization is the result of a reaction without any input of energy in which a **probe** attaches itself to a **target nucleic acid sequence** that is complementary to it. This target nuclear acid is present *in situ* in a tissue whose morphology must be preserved.

The probe is itself a nucleic acid whose sequence is known. This sequence is determined according to different criteria: **complementarity** and the fact of being **anti-sense** to the sequence of the target nucleic acid.

To detect the hybrid, the probe carries a label that is to be revealed.

↪ This phenomenon is identical for every instance of molecular hybridization. It is said to be specific when the hybrid formed is perfectly complementary.

↪ There exist different types of probe, but the principle is always the same.

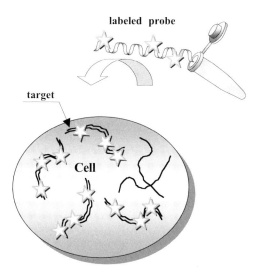

Figure 4.1 General diagram.

In situ hybridization is the step during which the target nucleic acid, having been rendered accessible to the probe contained in a hybridization solution, will be capable of hybridizing specifically in given conditions.

↪ The major disadvantage is that it is impossible to determine the molecular characteristics (e.g., molecular weight or sequence) of the nucleic acid revealed in the tissue.

Hybridization

With this approach, it is possible to localize the nucleic acid being sought, and to determine the tissue and cell structures in which it is contained (Figure 4.2).

The opposite of hybridization is denaturation, which happens at melting point, and which consists in the separation of the two strands of the hybrid.

↪ The advantage of this approach is that it can be used to detect the presence of a particular nucleic acid among all the constituents of a cell.

↪ Denaturation is generally carried out by raising the temperature.

↪ See Section 3.10

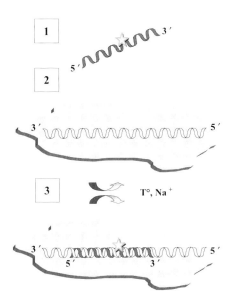

1 = Labeled probe.

2 = Target nucleic acid.

3 = Hybrid.

Figure 4.2 Principle of hybridization.

❑ *Nature of hybrids*

• Stable hybrids — Stable hybrids are characterized by the creation of all the hydrogen bonds involved in the formation of the new molecule (Figure 4.3).

↪ A perfect, very stable double strand is obtained.

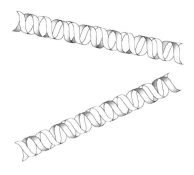

RNA probes give the most stable hybrids.

Figure 4.3 Stable hybrids.

↪ In increasing order of stability:
DNA – DNA < DNA – RNA < RNA – RNA

Hybrids that are very rich in G and C bases are much more stable than those that are rich in A and T/U bases.

• Unstable hybrids — Hybrids containing a lot of A and T/U bases are less stable, and thus easier to denature (Figure 4.4).

⇨ There are two hydrogen bonds between A and T (*see* Figure 4.13), but three between G and C (*see* Figure 4.14).

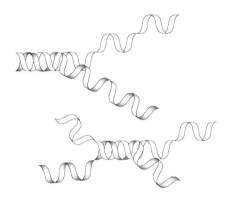

Figure 4.4 **Unstable hybrids.**

If the hybridization temperature is a lot lower than the melting temperature (Tm), many hybrids are formed, some of which are unstable.

⇨ *See* Section 4.2.3.1 (Determination of Tm).

• Partial hybrids — With low levels of homology of sequences between the two molecules of nucleic acid, matching will not be total (Figure 4.5).

⇨ These are mismatched hybrids, where only a certain number of hydrogen bonds are involved in the formation of the molecule (unstable hybrids).

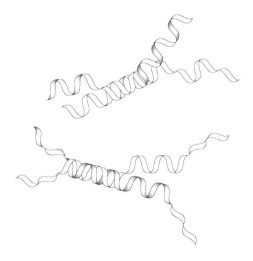

Figure 4.5 **Partial hybrids.**

4.2 PARAMETERS

The hybridization conditions must permit the formation of specific hybrids by limiting that of nonspecific bonds.

↪ However, it is often easier, or indeed necessary, to form all the hybrids by bringing about the constitution of nonspecific bonds (hybrid or probe–protein) which subsequently have to be eliminated.

The specificity of the hybridization process, and carrying it out *in situ*, depends on a number of parameters:

1. Hybridization temperature
2. Salinity
3. Hybridization buffer
4. Type of probe
5. Length of the probe

6. Type of label used
7. Homology of sequence: probe–target nucleic acid

↪ Depends on Tm
↪ Concentration of Na^+
↪ Ionic strength
↪ cDNA vs. oligoprobe vs. cRNA
↪ Stability of the hybrid dependent on the length of the probe
↪ Radioactive or antigenic labels

4.2.1 Structure of Nucleic Acids

4.2.1.1 Nucleoside

Nucleosides are molecules constituted by a puric or pyrimidic base and a sugar (ribose or deoxyribose).

1. The sugar (Figure 4.6).

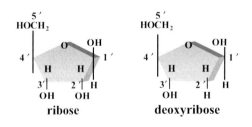

↪ Sugar with five carbons: the carbons in the sugar are numbered 1' to 5' to differentiate them from the carbons in the base cycle, which are numbered 1 to 7.
- Ribose in RNA
- Deoxyribose in DNA (absence of an atom of oxygen at the 2' carbon position)

Figure 4.6 Ribose and deoxyribose.

2. *The base (Figure 4.7).*

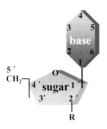

⇝ Base fixed by its carbon to the 1' carbon of the sugar:
- Puric bases (adenine and guanine)
- Pyrimidic bases (cytosine, thymine, or uracil)
- **R**: H (deoxyribonucleoside)
- **R**: OH (ribonucleoside)

Figure 4.7 Nucleoside: base–sugar combination.

4.2.1.2 Nucleotide

Triphosphate nucleotides are used as probe labels. Depending on the type of label (radioactive or antigenic), the nucleotides are modified by the addition of either a radioactive isotope or a molecule of the antigenic type.

❏ *Nucleotides*

Nucleotides are the elementary units of nucleic acids. A nucleotide is made up of three elements: (Figure 4.8)

1. *A puric or pyrimidic base*

2. *A sugar*
   - Deoxyribose
   - Ribose

3. ***One or more molecules of phosphoric acid*** fixed to the 5' carbon of the sugar

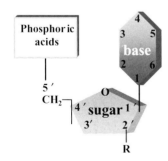

❏ *Phophodiester bonds (Figure 4.9)*

⇝ *See* Sections 1.2.1 and 1.2.2.

⇝ Nucleotide = nucleoside + phosphate group(s)

⇝ Grafting of an antigen-type molecule onto the nitrogenous base in the case of antigenic probes (*see* Section 1.2.1.2).

⇝ Substitution of one or more hydrogen atoms in the base cycle in the case of tritiated probes (*see* Section 1.2.1.1.4).

⇝ Deoxyribonucleotide (DNA)
⇝ Ribonucleotide (RNA)
⇝ A nucleotide triphosphate is made up of three phosphate groups. Depending on their position on the carbonated skeleton of the sugar, the phosphorus atoms are termed α, β, γ (*see* Section 1.1.2).

- **R**: H (deoxyribonucleoside)
- **R**: OH (ribonucleoside)

Figure 4.8 Nucleotide.

⇝ Nucleotides are linked together by phosphodiester bonds, the phosphate being situated between the 3' carbon of a given sugar and the 5' carbon of the following sugar. The synthesis is oriented in the 5' → 3' sense.

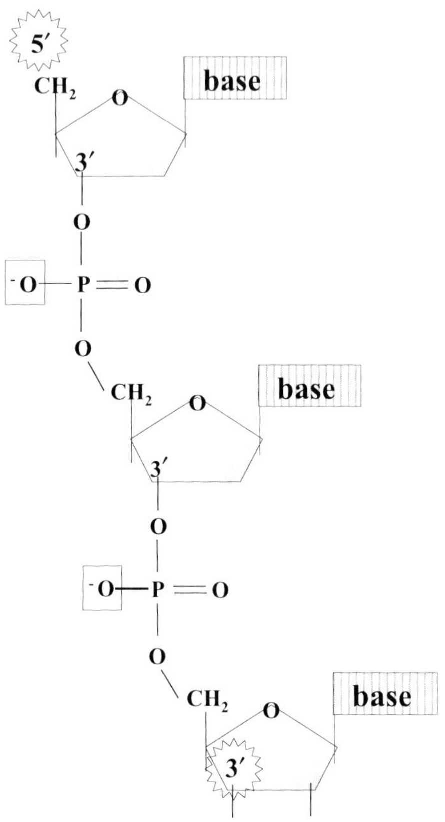

Hybridization

Figure. 4.9 Phosphodiester bonds.

4.2.1.3 DNA

Deoxyribonucleic acid is a double-stranded molecule composed of two **anti-parallel** nucleotide chains linked together by hydrogen bonds between the bases.

The four bases that make up DNA are: **adenine, guanine, cytosine** and **thymine.**

Each chain is made up of a sequence of **deoxyribonucleotides** linked together by phospho diester bonds which combine the 3' carbon of a given sugar and the 5' carbon of the following sugar. The nucleotide bond mobilizes two acid functions of the phosphoric group, with the remaining O⁻ function giving the polymer its neutralized acid characteristic *in vivo* by interaction with basic proteins (histones).

↬ Complementary chains whose 5' → 3' orientations are opposed (*see* Figure 4.10)

↬ Endogenous deoxyribonucleic acid is present in cell nuclei and mitochondria, and carries genetic information.

↬ dATP, dGTP, dCTP, and dTTP

↬ *See* Figure 4.9.

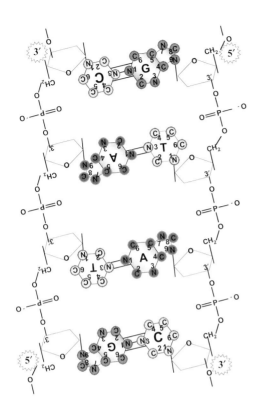

C= cytosine
G= guanine

A= adenine
T= thymine

Figure 4.10 Structure of DNA.

4.2.1.4 RNA

Ribonucleic acid is a single-stranded polynucleotide formed by the linking up of four structural ribonucleotides similar to DNA (Figure 4.11). The phosphodiester bridge brings together a given 3' carbon with the 5' carbon of the following ribonucleotide.

⇝ ATP, GTP, CTP, and UTP

The differences between RNA and DNA are the following:

1. RNA is single stranded.

⇝ DNA is double stranded.

2. The 4 bases that make up RNA are: **adenine, guanine, cytosine,** and **uracil**, linked together by phosphodiester bonds.

⇝ In RNA, uracil replaces thymine.

3. The sugar molecule in RNA is **ribose.**

⇝ Deoxyribose is the sugar molecule in DNA.

4. The RNA molecule is shorter than that of DNA.

Hybridization

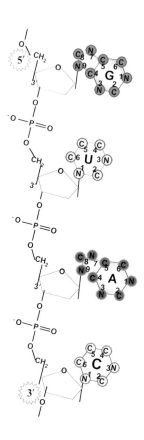

G= guanine

U= uracil

A= adenine

C= cytosine

Figure 4.11 Structure of RNA.

4.2.1.5 Duplication

In vivo, the DNA molecule is capable of making an exact copy of itself by a process called duplication.

4.2.1.6 Replication

During its synthesis, DNA is capable of making a replica of itself, or a very faithful copy. In the course of this process, the DNA double helix is unwound, and the two chains separate out. At the replication bifurcation, there is a synthesis of two new strands by complementary matching of the nucleotides that line each of the two strands.

Each DNA molecule thus possesses a new and an old strand.

➯ *In vitro*, the methods of labeling probes — random priming, nick translation — use this property (*see* Section 1.3).

➯ With bacteria, replication is very rapid: on the order of 750 to 1000 bp / s.
➯ In eukaryotic cells, it is much slower.

➯ This process is the basis of heredity.

4.2.1.7 Transcription

The DNA sequence that serves as the matrix is copied (transcribed) in sequences of RNA, giving a single-strand RNA that is complementary to the matricial DNA sequence.

This process is catalyzed by the RNA polymerase enzyme, which copies DNA strands only in the 5' → 3' sense. The RNA chain grows in the 3' → 5' sense.

Transcription produces several types of RNA:
- Messenger RNA, which carries the information required for the synthesis of proteins
- Transfer RNA (Figure 4.12), which transports amino acids during the synthesis of proteins

- Ribosomic RNA

↪ This is the process called *in vitro* transcription (*see* Section 1.3.8).

↪ RNA polymerase recognizes errors resulting from incorrect matching. Its exonuclease activity consists of replacing wrong nucleotides with the appropriate nucleotides.

↪ Molecules of between 150 and 12,000 nucleotides
↪ Molecules of between 73 and 93 nucleotides
↪ Partially double-stranded structure
↪ This molecule has a clover-leaf shape. The anticodon is positioned at one end of the molecule, while the 3' end, which always terminates with a CCA sequence, corresponds to the fixation site of the activated amino acids.

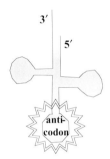

Figure 4.12 Transfer RNA.

↪ Components of ribosomes

4.2.2 Complementarity of Bases

There exist five principal bases, subdivided into two families:

- **The puric bases**
— Guanine and adenine, in both DNA and RNA
- **The pyrimidic bases**
— Cytosine and thymine (DNA)
— Cytosine and uracil (RNA)

❏ *Matching of bases*
Hybridization between the probe and the target nucleic acid takes place through the formation of hydrogen bonds between its bases:

↪ Adenine, guanine, cytosine, thymine, and uracil

↪ The purines

↪ The pyrimidines

↪ There is a complementarity between the bases: a pyrimidic base is always associated with a puric base.

Hybridization

1. *Matching of the adenine and thymine/uracil bases (Figure 4.13).*

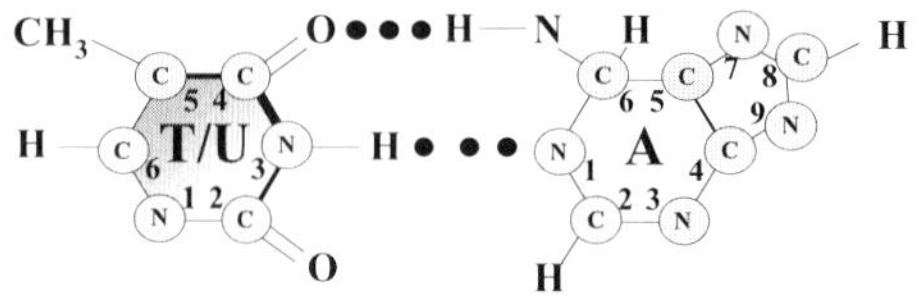

⇝ A – T / U base groups are linked by two hydrogen bonds:
• In DNA–DNA hybrids, the puric base adenine is complementary to the pyrimidic base thymine.
• In DNA–RNA hybrids, the puric base adenine is complementary to the pyrimidic base uracil.

Figure 4.13 A – T / U matching.

2. *Matching of the cytosine and guanine bases (Figure 4.14).*

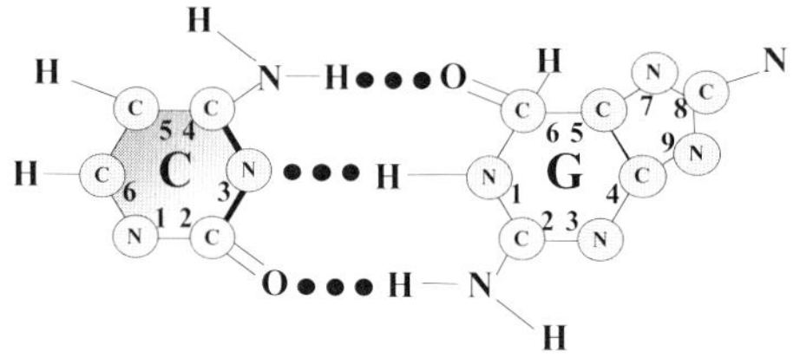

⇝ C – G base groups are linked together by three hydrogen bonds.
⇝ Whatever the hybrid, the puric base guanine is always linked to the pyrimidic base cytosine.

Figure 4.14 C – G matching.

Hybrids that are rich in C – G bases are much more stable than those that are rich in A – T / U bases.

⇝ The presence of too many C – G bases brings about nonspecific combinations. But on the other hand, hybrids containing a large number of A – T / U bases are very easy to denature.

4.2.3 Hybridization Temperature

This is the temperature at which hybrids form, which is what determines the specificity of a hybrid. The way in which temperature influences the formation of hybrids is schematized in Figure 4.15.

⇝ The formation of hybrids is a spontaneous reaction that can be modulated by the hybridization temperature.

4.2.3.1 Determination of Tm

⇝ Tm is the temperature at which 50% of the complementary strands form hybrids, and 50% are denatured.

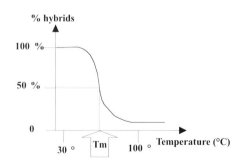

Figure 4.15 Determination of Tm.

The temperature that corresponds to 50% of hybrids is that of fusion, or Tm.

↪ This temperature corresponds to the point of inflection of the sigmoid curve.

Tm is a constant for a given hybrid.

↪ A rise in temperature causes the denaturation of hybrids. The curve that describes this denaturation is a sigmoid curve.

For each type of probe, numerous formulae are available in the literature.

↪ Probes include cDNA, oligonucleotide, and cRNA probes.

1. *cDNA probe* —The formula is:

   a. For large cDNA probes:
   $Tm_{DNA} = 16.5 \log (Na^+) + 0.41 (\% C + G) + 81.5$

   ↪ >500 nucleotides. This formula was determined experimentally, and is used to calculate the Tm as a function of the **C–G** concentration, according to the **saline concentration**.

   b. For smaller cDNA probes:
   Tm_{DNA} = long duplex Tm − 500/number of short duplex bases

   ↪ Between 100 and 300 nucleotides. This length corresponds to the probe fragments after labeling.

2. *Oligonucleotide probe*
   a. There are also formulae for short probes, notably for oligonucleotide probes:

   ↪ <18 nucleotides

   $Tm_{oligo} = 4 [\% C + G] + 2 [\% A + T]$

   ↪ This formula takes account only of the number of bases in the sequence, and not the salinity.

   b. For longer oligonucleotide probes, Tm is generally calculated at the time of synthesis.

   ↪ <18 nucleotides

3. **RNA probe**
   $Tm_{RNA} = 18.5 \log [Na^+] + 58.4 [\% G + C] - 0.72 [\% \text{formamide}] + 79.8 - 820 / \text{number of bp in the duplex}$

   ↪ RNA–DNA duplex

4.2.3.2 Determination of the hybridization temperature

Empirically, the hybridization temperature is determined as a function of the Tm, and of the nature of the probe:

↪ It must be lower than the Tm, to obtain more than 50% of hybrids from two complementary strands.

1. *cDNA probe* — 10°C below Tm

   ↪ 200 to 300 bp

2. *Oligonucleotide probe* — ≈ 5°C below Tm

   ↪ 20 to 30 mers

3. *cRNA probe* — ≈ 15°C below Tm

   ↪ 200 to 300 mers

Hybridization

If the hybridization temperature is higher than the Tm, it favors thermal agitation, and therefore the instability of hybrids, which can lead to fusion.

The closer the hybridization temperature is to the Tm, the more specific is the hybridization, since the hybrids are stable. The probability of obtaining combinations of weak homologies is thereby reduced.

⇝ Its disadvantage lies in the formation of only a small number of hybrids.

If the hybridization temperature is a lot lower than the Tm, numerous hybrids form, and some of these are unstable, or of low affinity.

⇝ This is the most widely used approach. Unstable hybrids are eliminated during the washing steps. Its advantage is that it gives a maximum of hybrids.

This temperature can be artificially raised by the addition of formamide.

⇝ The addition of 1% of formamide lowers the hybridization temperature by 0.65°C for DNA, and 0.72°C for RNA.

4.2.4 Na⁺ Concentration

⇝ Salinity

The monovalent Na⁺ ion concentration influences the stability of the hybrids. An increase of the order of 10 in this concentration brings about an increase of 16.6°C in the Tm. The consequence of this stabilization of the hybrids is an increase in hybridization speed.

⇝ The concentration that is generally used is 660 mM of Na⁺, which corresponds to an SSC 4X buffer (*see* Appendix B3.4).

4.2.5 Hybridization Buffer

Large variations in pH destabilize hybrids.

⇝ This parameter can limit interbase hydrogen bonds.

4.2.6 Type of Probes

DNA – DNA hybrids are less stable than DNA–RNA hybrids, which in turn are less stable than RNA – RNA hybrids.

⇝ RNA probes make it possible to obtain more stable hybrids.

4.2.7 Length of Probes

The length of the probe is a parameter that acts on the Tm, and thus on the stability of the hybrids.

⇝ The longer the probe, the more stable the hybrid. On the other hand, its penetration of the tissue will be limited.

4.2.8 Type of Label Used

The label causes spatial and stereospecific modifications of nucleotides, with the exception of the ^{32}P, ^{33}P, and ^{3}H labels, which are an integral part of the molecule.

The 3' extension of an oligonucleotide can result in a labeled oligonucleotide two to three times longer than the oligonucleotide to be labeled.

The hybridization temperature should be inversely proportional to the number of antigenic nucleotides incorporated.

⇝ *See* Figures 1.4, 1.5, and 1.7 through 1.9.

⇝ The substitution of an oxygen atom by a ^{35}S causes less disturbance than the modification of a base by the addition of an antigen.

⇝ The addition of too many radioactive nucleotides in extension is detrimental to hybrid formation.

⇝ With the addition of a carbonated chain and an antigenic molecule, antigenic labels modify the ability to form hydrogen bonds, and thus the behavior of the hybrid. Labels in the 3' extension are those that cause least modification to the structure of the probe.

4.2.9 Homology of Sequence

The aim of hybridization is to detect a nucleic acid whose sequence is complementary to that of the probe. However, a percentage of heterologous bases may be acceptable (1% of mismatches reduces the temperature of fusion by around 1°C).

⇝ The degree of homology between the two molecules of nucleic acid determines the stability of the hybrid.

⇝ The positions of the mismatches are important. Those that are situated at the extremities are less involved in the stability of the hybrid.

4.3 POSTHYBRIDIZATION TREATMENTS

4.3.1 Purpose of Washing

The specificity and stability of the matching between two molecules of nucleic acid depend on their degree of complementarity.

⇝ Thus, during the hybridization step, the probe fixes with a very high level of affinity to perfectly homologous nucleic acid molecules (**matched hybrid**) (*see* Figure 4.3), but also, with a lower level of affinity, to nucleic sequences that present only partial homology (**mismatched hybrid**) (*see* Figure 4.5).

Hybridization

The purpose of washing is to eliminate nonspecific hybridization, which introduces background and masks specific hybridization by reducing the signal/noise ratio. A succession of washings is carried out in increasingly stringent conditions such that only specific hybrids, which are the most stable, persist.

↪ Furthermore, a probe tends to fix in a totally aspecific way to non-nucleic structures, essentially those of proteins.

4.3.2 Stringency Conditions

The parameters that determine the stringency conditions, and thus the choice of hybrids according to their stability, are:

1. *Temperature*

2. *Ionic strength*

3. *Washing time*

The tighter the stringency conditions, the more the hybrids are denatured.

↪ Mismatched hybrids are denatured before matched hybrids.

4.3.2.1 Determination of the Tw

The higher the temperature, the higher the risk of denaturation.

1. The washing temperature (Tw) is obtained by doing a washing bath with SSC 2X at temperatures varying by 5°C between 40°C and 70°C (*see* Figure 4.16).

2. The density of the hybridization signal is determined by the optical measurement of the optical density of the macroautoradiograms (Figure 4.16). The higher the washing temperature, the more the hybrids dissociate. At Tm, 50% of specific hybrids are denatured; beyond this point, the majority of hybrids are denatured. At 70°C, all the DNA – DNA and RNA – DNA hybrids are dissociated.

↪ Tw: washing temperature.

↪ Water baths with shaking give the best reproducibility of results.
↪ The use of a radioactive probe facilitates the quantification of the signal.

↪ The shape of this washing curve informs one about the nature of the hybrids formed. A sigmoid curve is typical of a specific hybrid.

↪ The aim is to eliminate the maximum of nonspecific hybrids, while preserving the maximum of specific hybrids.

⇨ Only a sigmoid curve allows the denaturation of a hybrid to be detected. Any other curve will leave open the question of a nonspecific hybridization:

- A straight line is typical of an aspecific bond

- The sigmoid + straight line combination is the general case, with specific hybrids but aspecific bonds

- Several consecutive sigmoids constitute the expression of a heterologous population of hybrids

- A high level of signal at more than 70°C is characteristic of an aspecific bond, which requires complementary pretreatments.

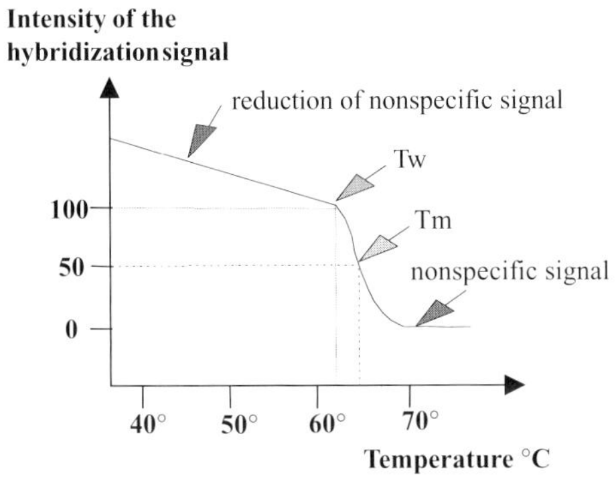

Figure 4.16 Washing curve for a hybridization (Tw).

The Tw is determined by the sigmoid part of the washing curve.

⇨ This washing temperature varies from one probe to another, and, for a given probe, from one type of tissue to another.

4.3.2.2 Ionic strength

The salt (Na^+) concentration is inversely proportional to the stability of the hybrids. Washing in decreasing concentrations of Na^+ brings about the dissociation of the hybrids, according to their stability: the less stable are denatured before the more stable.

⇨ Mismatched hybrids and nucleic acid–protein bonds are less stable than specific hybrids.
⇨ Washing in water should cause total denaturation.

4.3.2.3 Washing time

Little difference is observable after an hour.

⇨ Washing accompanied by intense or prolonged shaking can cause damage to sections, and is more difficult to reproduce.

Ceteris paribus, repeated washings are more effective than long washes.

⇨ The repetition of washes can be compensated for by voluminous washing and shaking.

Hybridization

4.4 SUMMARY OF THE DIFFERENT STEPS

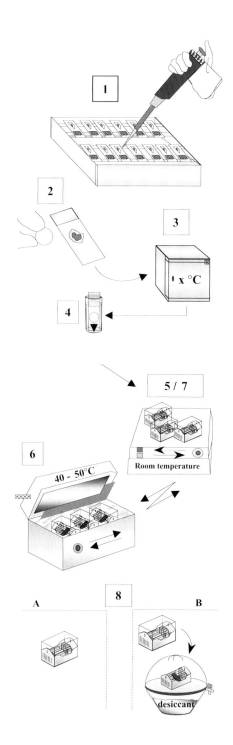

1 = **Placing the probe.**

2 = **Placing the coverslip.**

3 = **Hybridization** at the appropriate temperature.

4 to 7 = **Rinsing** in decreasing concentrations of SSC:
- Removal of the coverslip (4)
- At room temperature (4 at 5)
- At Tw (6)
- At room temperature (7).

Enzymatic treatment

8 = **Revelation**
A = **Antigenic probe** — immunological revelation (*see* Chapter 5)

B = **Radioactive probe:**
- Dehydration
- Drying
- Autoradiographic revelation (*see* Chapter 5)

Figure 4.17 Summary of the different hybridization steps.

4.5 EQUIPMENT/REAGENTS/SOLUTIONS

4.5.1 Equipment

- Centrifuge
- Coverslip
  — 12 × 12 mm coverslip
  — 24 × 50 mm coverslip
  — ∅ 9 mm circular coverslip
- Gloves
- Moisture chamber

- Ovens
- Shaker
- Slide tweezers

- Staining trays + slide-holder
- Surseal
- Vacuum jar or Speedvac
- Vortex
- Water bath

↝ Outside of the hybridization process itself, sterilization of the equipment is not indispensable.
↝ >14,000 g
↝ Sterilized coverslips, wrapped in aluminum foil and stored at room temperature; gloves have to be worn when handling them.

↝ Of the petri box type (≈ 24 × 24 cm), with enough space for the slides
↝ 37 to 65°C
↝ Low-amplitude horizontal shaking
↝ Sterilized tweezers, wrapped in aluminum foil and stored at room temperature
↝ Conventional equipment
↝ Can be replaced by rubber cement
↝ Only with autoradiographic revelation

↝ Horizontal shaking

4.5.2 Reagents

- Dextran sulfate
- Dithiothreitiol (DTT) (Figure 4.18)

↝ AR quality, to be used only for *in situ* hybridization

↝ Also exists in solution
↝ Molecular biology quality
↝ $C_4H_{10}O_2S_2$

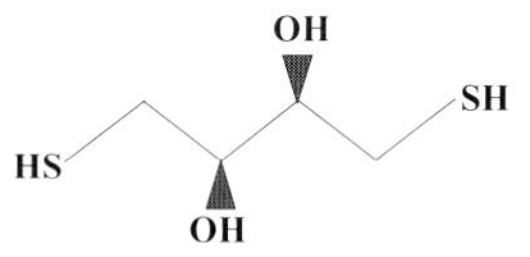

Fig. 4.18 Dithiothreitiol (DTT)

- DNA
- Formamide
- Heparin
- Poly A
- RNA
- RNase A
- Rubber cement
- Sarcosyl
- Silica gel
- Sodium chloride
- Trisodium citrate

↝ Mainly salmon sperm DNA
↝ Exists in deionized form
↝ Injectable type
↝ Molecular biology quality
↝ Mainly yeast transfer RNA
↝ Destruction of all the RNA
↝ In tubes, for preference
↝ Detergent
↝ Desiccant, can be regenerated
↝ AR type, anhydride
↝ AR type, anhydride

4.5.3 Solutions

- Alcohol 70%, 95%, and 100% ↪ *See* Appendix B4.2.

- Deionized formamide ↪ *See* Appendix B2.13. Storage at –20°C in aliquots of 500 µL. If the reagent is not frozen at –20°C, it should not be used.

- Denhardt's solution 50X ↪ *See* Appendix B2.9. Storage at –20°C in aliquots. Thawing several times is possible without modification of properties.

- Dextran sulfate 50% ↪ *See* Appendix B2.22. Storage at –20°C

- Dithiothreitiol 1 M ↪ *See* Appendix B2.10. Storage at –20°C in aliquots of 50 to 100 µL. It cannot be refrozen. Its nauseating odor is the best guarantee of its condition. It should not be used if the odor is modified or absent.

- DNA 10 mg/mL ↪ *See* Appendix B2.3. Sonication. Storage at –20°C. The freezing/thawing cycles can lead to breaks.

- Heparin 0.5 mg/mL ↪ Storage at –20°C.

- Poly A 10 mg/mL ↪ *See* Appendix B2.14. Equivalent to RNA.

- RNA 10 mg/mL ↪ *See* Appendix B2.4. Storage at –20°C. The freezing/thawing cycles can lead to breaks.

- RNase A 100 µg/mL TE/NaCl ↪ *See* Appendix B2.18.

- Sarcosyl ↪ *See* Appendix B2.20.

- SSC 20X, pH 7.0 ↪ *See* Appendix B3.4. Storage at 4°C or at room temperature.

- Sterilized water ↪ *See* Appendix B1.1. Use only at the time of opening the bottle, or use DEPC-treated water (*see* Appendix B1).

- TE buffer/NaCl ↪ *See* Appendix B3.5.2.

4.6 COMPOSITION OF THE HYBRIDIZATION MEDIUM

↪ The hybridization buffer can be stored, at least for a few weeks, at –20°C, either alone or added to the probe, with no decrease in its effectiveness.

Solutions	Final Conc.
• SSC buffer 20X	4X

↪ Hybrids form at pH close to neutrality in the presence of Na^+ ions (ionic strength).
↪ The Na^+ ion concentration is the major factor in the stability of the hybrid, and has a direct effect on its stability.
↪ **Indispensable**

• Deionized formamide	50%

↪ The addition of formamide reduces the hybridization temperature (*see* Section 4.2.3.2).
↪ This increases the effectiveness of the hybridization process by reducing the temperature necessary for hybridization to take place, which gives better conservation of tissue morphology.
↪ **Useful.** The formamide must be of good quality (very pure) and deionized (for the maintenance of the pH of the buffer).

• 50% dextran sulfate	10%

↪ This increases the effectiveness of hybridization. A nonionic polymer, it concentrates the solute through the volume that it occupies in the solution, and accelerates the reaction.
↪ This protects cellular RNA.
↪ **Very useful** with cDNA probes.

or

• Heparin 0.5 mg / mL	0.1 mg/mL
• RNA 10 mg / mL	100 to 250 µg/mL

↪ This can be substituted for dextran sulfate.
↪ This produces saturation of nonspecific bonds of the proteic type by competition, hence its high concentration.
↪ **Useful.**

• DNA 10 mg/ml	100 to 400 µg/mL

↪ This produces saturation of nonspecific bonds.
↪ **Useful** with cDNA probes.

• Poly A 10 mg/ml	100 µg/mL

↪ This produces saturation of nonspecific bonds. It can be replaced by a poly C or a poly G.

• Denhardt's solution 50X	1 to 2X

↪ This produces saturation of nonspecific bonds by competition with macromolecules. Its concentration can be increased.
↪ **Useful.**

• 1 *M* DTT	10 m*M*

↪ This is to be used only with probes labeled with ^{35}S (antioxidant against the radiolysis of S^{35}).
↪ **Important:** This is to be used only after denaturation, since it is destroyed at 100°C.

4.7 PROTOCOLS

4.7.1 Protocol for cDNA Probe

4.7.1.1 Hybridization Buffer

↪ It can either be prepared at time of use or conserved, either at –20°C or –80°C.

Solutions	Final Conc.	
1. Add		
• SSC buffer 20X, pH 7.0	4X	↪ **Indispensable**
• Deionized formamide	50%	↪ **Useful**
• 50% dextran sulfate	10%	↪ **Useful**
• RNA 10 mg / mL	100 to 250 µg/mL	↪ **Useful**
• DNA 10 mg / mL	100 to 400 µg/mL	↪ **Useful**
• Poly A 10 mg / mL	100 µg/mL	
• Denhardt's solution 50X	1 to 2X	↪ **Useful**
• 1 M DTT	10 mM	↪ Only for probes labeled with ^{35}S
2. Vortex.		
3. Centrifuge.		

4.7.1.2 Reactive medium

1. Add the probe to the hybridization buffer at the "saturating" concentration:

↪ The addition of any higher concentration of probe will not modify the intensity of the signal.
↪ To check this saturating concentration, incubate adjacent slides with increasing concentrations of probe. Beyond a certain concentration, the signal no longer increases, and the first concentration that gives a signal is the saturating concentration. In practice, the probe is labeled with a radioactive isotope, and the density of the film is measured by densitometry.

- Radioactive probe   0.1 to 0.2 µg/mL hybridization buffer

↪ After centrifugation and drying, the labeled probe is dissolved directly in the hybridization buffer. Its concentration depends on the nature of the label (*see* Section 1.5.1).

- Antigenic probe   1 to 10 µg/mL hybridization buffer

↪ *See* Section 1.5.2.

2. Vortex.
3. Centrifuge.
4. Denature.

    3 min at 100°C

↪ *See* Figure 4.19.
↪ **Indispensable.** Double-stranded cDNA probes must be denatured.

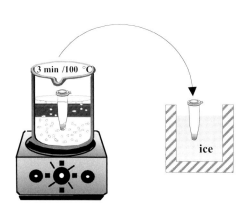

↪ The tube containing the mixture is placed on a floater at the surface of a water bath at 100°C for 3 min to denature the DNA. The cap should be pierced if the volume is >50 μL; otherwise a clip can be used.

Figure 4.19 Denaturation.

5. Cool rapidly in ice.

↪ This is done to avoid the rehybridization of two denatured strands. At 0°C, single strands are stable.

6. Add DTT.   **X μL**

↪ This is done only for ^{35}S probes.
↪ Final concentration: 10 mM

4.7.1.3 Hybridization

1. Place a suitable surseal around the section.

2. Place the hybridization solution on the sections, either pretreated or not.

   - Circular coverslip   **≥10 μL**
   - 12 × 12 mm coverslip   **≥15 μL**
   - 24 × 50 mm coverslip   **≥50 μL**

3. Cover with a sterile coverslip.

4. Place the slides flat in a moisture chamber.

5. Incubate.   **Overnight at 40 to 42°C**

↪ **Gloves should be worn**
↪ Optional (*see* Step 4).
↪ Or seal the coverslip with rubber cement
↪ The amount of hybridization solution to be used, and the size of the coverslip, vary according to the size of the section.

↪ Avoid air bubbles and make sure that the slide is properly horizontal.
↪ The concentration of liquid in the moisture chamber must produce a vapor saturation identical to that of the hybridization buffer: several layers of filter paper soaked in SSC 5X should be placed at the bottom of the box. In this way, Step 1 can be avoided.
↪ **Use moisture chamber.**
↪ *See* Section 4.2.3.
↪ These are the temperatures that are generally used to obtain all the hybrids, with the aim of eliminating nonspecific hybrids during washes.
↪ This duration is generally used because it is practical, but it is not obligatory. Shorter times, i.e., a few hours, give identical results.

4.7.1.4 Washing

- Duration of washing steps
  ⇒ Allow 5 h.
- Amount of buffer necessary
  ⇒ Allow a minimum volume of 200 mL for each rinse (tray of 20 slides).
  ⇒ Rinses are more numerous with radioactive probes.
- Conditions of manipulation — Wearing gloves is not obligatory
  ⇒ The presence of RNase is not a danger, because only single-stranded RNA can be destroyed by this enzyme. But all the hybrids are by definition double stranded, and so they cannot be broken down by it.
- Rinses carried out in SSC buffer, either at room temperature, with slow shaking, or at Tw in a water bath with shaking.
  ⇒ The SSC buffer is at pH 7.0. A basic pH can cause partial denaturation of the hybrids.

❏ *Protocol*

• SSC 5X	5 min	⇒ The first rinse serves only to unstick the slides: place the slides in a borrel tube or tray containing the solution; let the slide slide gently without touching it (dissociation).
• SSC 2X	1 h Tw	⇒ **Necessary.** ⇒ For washing temperatures (*see* Section 4.3.2)
• SSC 2X	2 × 30 min	
• SSC 1X	1 h	
• SSC 0.5X	1 h	⇒ **Optional step**. Temperature is a factor of stringency, which at this concentration of Na^+, is very effective. Use a water bath with shaking.
• SSC 0.1X	30 to 60 min	⇒ **Optional step**. This is to be used only with radioactive probes. The Na^+ concentration is very low, and can cause dissociation of specific hybrids.

Following step — Before revelation.
⇒ See Section 4.8.

4.7.2 Protocol for Oligonucleotide Probe

⇒ **Gloves should be worn.**

4.7.2.1 Hybridization Buffer

⇒ It can be prepared at time of use, or stored for some weeks at $-20°C$ or $-80°C$

Solutions	Final Conc.	
1. Add:		
• SSC buffer 20X	4X	⇒ **Indispensable**
• Deionized formamide	50%	⇒ **Useful**
• Dextran sulfate 50%	10%	⇒ **Useful**
• tRNA 10 mg / mL	250 µg/mL	⇒ **Indispensable**

• DNA 10 mg / mL	100 to 400 μg / mL	↝ Useful
• Sarcosyl	0.8X	↝ Optional
• Denhardt's solution 50X	2X	↝ Useful
• DTT 1 M	10 mM	↝ Only for ^{35}S probes ↝ Antioxidant against the radiolysis of ^{35}S

2. Vortex.   ↝ Slowly

3. Centrifuge.   ↝ To collect all the drops

4.7.2.2 Reactive medium

1. In an Eppendorf tube, resuspend the probe in the hybridization buffer:

↝ **Gloves should be worn**
↝ After centrifugation, the labeled probe is dissolved directly in the hybridization buffer.
↝ This concentration depends on the nature of the label.

• Radioactive probe	2 to 5 pM / mL hybridization buffer	↝ *See* Section 1.6
• Antigenic probe	10 to 200 pM / mL hybridization buffer	↝ The probe remains stable for several months at –20°C (*see* Section 1.6)

2. Vortex — No denaturation of the probe is necessary.

↝ Single-stranded probe

4.7.2.3 Hybridization

↝ **Gloves should be worn.**

1. Place a suitable surseal around the section.

↝ Optional (*see* Step 4 in Section 4.7.1.3).
↝ Or seal the coverslip with rubber cement

2. Place the hybridization solution on the sections, either pretreated or not.

↝ *See* Step 1 in Section 4.7.1.3).

3. Cover with a sterile slide.

↝ Avoiding air bubbles, make sure that the slide is properly horizontal.

4. Arrange the slides horizontally in a moisture chamber.

↝ See Section 4.7.1.3, Step 4.
↝ SSC buffer 5X.

5. Incubate.   **3 h to overnight 35 to 45°C**

↝ Use a **moisture chamber.**
↝ Hybridization temperature is defined as a function of the probe characteristics (*see* Section 4.2.3).
↝ Temperatures in the range of 35 to 45°C are generally used to obtain all the specific hybrids, with the aim of eliminating nonspecific hybrids during the washes.

Hybridization

4.7.2.4 Washing

- Washing time
- Amount required

- Conditions of manipulation — The wearing of gloves is not obligatory.

- Rinsing is carried out in SSC buffer at pH 7.0, either at room temperature with slow shaking or at Tw in a water bath with shaking.

❏ *Protocol*
- SSC 4X   5 min
- SSC 2X   1 to 2 h, at Tw
- SSC 2X   1 h
- SSC 1X   1 h
- SSC 0.5X   1 h

- SSC 0.1X   30 to 60 min

Following step — Before revelation.

↬ Gloves should be worn.

↬ Allow 6 h.
↬ Allow a minimum volume of 200 mL for each rinse (tray of 20 slides).
↬ The rinses are more numerous with radioactive probes.
↬ *See* Section 4.7.1.4 (cDNA probe washes).
↬ Gloves must be worn for radioactive-labeled probes.

↬ *See* Section 4.3.2.1 (for the determination of the washing temperature Tw, *see* Figure 4.16).

↬ **Necessary**
↬ *See* Section 4.3.2.
↬ **Necessary**
↬ **Necessary**
↬ **Optional step**. Temperature is a factor in stringency which, at this Na^+ concentration, is very effective. Use a water bath with shaking.
↬ See Chapter 7.
↬ **Optional step**
↬ See Chapter 7.
↬ See Section 4.8.

4.7.3 Protocol for cRNA Probe

4.7.3.1 Hybridization buffer

Solutions	Final Conc.

1. Add:
- SSC buffer 20X   4X
- Formamide deionized   50%
- Dextran sulfate 50%   10%
- RNA 10 mg / mL   100 to 250 µg / mL
- DNA 10 mg / mL   100 to 400 µg / mL
- Poly A 10 mg / mL   100 µg / mL
- Denhardt's solution 50X   1 to 2X
- 1 *M* DTT   10 m*M*

2. Vortex.

3. Centrifuge.

↬ Gloves should be worn.

↬ It can be prepared at time of use, or stored at –20 / –80°C for some weeks.

↬ **Indispensable**
↬ **Indispensable**
↬ **Necessary**
↬ **Indispensable**
↬ **Necessary**
↬ **Useful**
↬ **Necessary**
↬ Only for single-stranded ^{35}S probes

↬ Slowly

↬ Necessary that all the hybridization buffer be at the bottom of the tube

4.7.3.2 Reactive medium

1. Resuspend the labeled probe in an Eppendorf tube in the hybridization buffer:
- Radioactive probe   **0.01 to 0.1 µg / mL hybridization buffer**
- Antigenic probe   **0.1 to 1 µg / mL hybridization buffer**
2. Vortex — cRNA probes generally do not need the denaturation step.

↪ **Gloves should be worn**
↪ *See* Section 4.7.1 (cDNA probe protocol).
↪ *See* Section 1.6.1.

↪ *See* Section 1.6.1.

↪ Check that there are no secondary structures.

4.7.3.3 Hybridization

1. Place a suitable surseal around the section.

2. Place the hybridization solution on the sections, pretreated or not.
3. Cover with a sterile slide.

4. Arrange the slides in a horizontal position in a moisture chamber.
5. Incubate.   **Overnight 40 to 60°C**

↪ **Gloves should be worn.**
↪ Optional (*see* Section 4.7.1.3, Step 4).
↪ Or seal the coverslip with rubber cement.
↪ *See* Section 4.7.1.3, Step 4.

↪ Avoiding air bubbles, make sure that the slide is properly horizontal.
↪ *See* Section 4.7.1.3, Step 4.
↪ SSC buffer 5X.
↪ The temperature is always higher than for probes of the cDNA or oligonucleotide type. The temperature is defined in terms of the characteristics of the probe (*see* Section 4.2.3).
↪ Use a moisture chamber.

4.7.3.4 Washing

- Duration of washing
- Amount required

- Conditions of manipulation — The wearing of gloves is not obligatory

- Rinsing is done in SSC buffer, either at room temperature with slow shaking or at Tw in a water bath with shaking.
❑ *Protocol*
- SSC 4X   **5 min**

- SSC 2X   **1 h at Tw**

↪ **Gloves should be worn.**
↪ Allow ≈ 7 h.
↪ Allow a minimum volume of 200 mL for each rinse (tray of 20 slides).
↪ There are more rinses with radioactive probes.
↪ The use of RNases is recommended for single-stranded RNA; hybrids, being double stranded, are not broken down, although a non-hybridized probe would be.
↪ *See* Section 4.3.2 (for the determination of the washing temperature Tw, *see* Figure 4.16).

↪ *See* Section 4.7.1 (washing for cDNA probe).

↪ **Indispensable.** Because the RNA – RNA and RNA – DNA hybrids are stable, the washing times are longer.

Hybridization

• SSC 2X	2×30 min	
• SSC 1X	1 h	
• TE/NaCl	30 min at 37°C	↝ Put the TE/NaCl in an oven at 37°C for 30 min before adding the RNase.
• RNase A treatment — 100 µg/mL TE/NaCl	20 to 30 µg/mL 45 to 60 min at 37°C	↝ This step is **necessary**, and preferable to rinsing at high temperatures, which risks degrading the sections. RNase A allows non-hybridized cRNA (in the form of single strands) to be broken down, associated with proteins, or other macromolecules. Its effectiveness is visible in the intensity of the background noise.
• TE/NaCl	3×10 min	↝ This buffer, which is very stable, effectively inhibits the action of the enzymes that degrade DNA. Sodium chloride stabilizes hydrogen bonds that exist between matched bases, but also makes the DNA molecule more rigid.
• SSC 0.5X	1 rapid bath	↝ This is the buffer change step.
• SSC 0.5X	1 h at 45°C	↝ **A step that is often necessary.** The combination of temperature and reduction of Na^+ concentration increases stringency. ↝ It is possible to increase this to reduce background noise (*see* Chapter 7).
• SSC 0.5X	30 min	↝ **A step that is often necessary**
• SSC 0.1X	30 to 60 min	↝ **A step that is often necessary**, which is to be used only with radioactive probes. The Na^+ concentration is very low, and may cause a dissociation of specific hybrids.
Following step — Before revelation.		↝ *See* Section 4. 8.

4.8 BEFORE REVELATION

According to the label:
- Radioactive label
- Antigenic label

Two revelation methods are possible.

↝ The probe label is at this stage the hybrid label.

4.8.1 Radioactive Label

4.8.1.1 Dehydration
- Alcohol 70% **1 rapid**

- Alcohol 95% **1 rapid**
- Alcohol 100% **1 rapid**

4.8.1.2 Drying
- In a vacuum jar **30 min to 1 h**

Following steps:
- Contact with autoradiographic film
- Dip in the autoradiographic emulsion

⇝ The sections have to be dried before contact with autoradiographic film (*see* Section 5.1.3.5).

⇝ It is advisable to make up the solutions of alcohol with 300 mM ammonium acetate to maintain the ionic strength. As this salt is volatile, it does not cause deposits during the drying of the sections.

⇝ Gently wipe the edges of the sections and the backs of the slides.

⇝ **Indispensable**
⇝ The aim of this step is to eliminate all traces of alcohol and water (especially for the tissues), which could cause background noise with macroautoradiograms done on film.

⇝ *See* Section 5.1.3.5.
⇝ *See* Section 5.1.4.5.

4.8.2 Antigenic Label

For immunohistological revelations, **never dry tissue sections**.
The tissue is transferred to the buffer that will be used during revelation.
Following step — Immunological revelation.

⇝ **Indispensable**

⇝ *See* Section 5.2.3.

Chapter 5

Revelation of Hybrids

Contents

Choice of Method			143
5.1	Radioactive Hybrids		143
	5.1.1	Principle	144
	5.1.2	Characteristics	145
		5.1.2.1 Radiation	145
		5.1.2.2 Photographic Emulsions	145
		5.1.2.3 Exposure	145
		5.1.2.4 Development	145
		5.1.2.5 Efficiency	146
		5.1.2.6 Localization *In Situ*	146
		5.1.2.7 Detection	146
		5.1.2.8 Artifacts	147
		5.1.2.9 Quantification	147
	5.1.3	Macroautoradiography	147
		5.1.3.1 Principle	147
		5.1.3.2 Choice of Autoradiographic Film	147
		5.1.3.3 Summary of the Different Steps	148
		5.1.3.4 Equipment/Reagents/Solutions	149
		5.1.3.5 Protocol	150
	5.1.4	Microautoradiography	151
		5.1.4.1 Principle	151
		5.1.4.2 Choice of Autoradiographic Emulsion	151
		5.1.4.3 Summary of the Different Steps	152
		5.1.4.4 Equipment/Reagents/Solutions	153
		5.1.4.5 Coating the Slides	154
		5.1.4.6 Microautoradiographic Revelation	155
5.2	Antigenic Hybrids		156
	5.2.1	Immunohistological Reaction	157
		5.2.1.1 Immunocytological Tools	157
		5.2.1.1.1 Antibodies	157
		5.2.1.1.2 Streptavidin	157
		5.2.1.1.3 Biotin	158
		5.2.1.2 Labels	158
		5.2.1.2.1 Enzymes	158
		5.2.1.2.2 Fluorescent Labels	159
		5.2.1.2.3 Particle Label	163
		5.2.1.2.4 Choice of Label	164
		5.2.1.2.5 Principle of Obtaining Conjugates	165
	5.2.2	Detection	167
		5.2.2.1 Fluorescence	167
		5.2.2.1.1 Principle	167
		5.2.2.1.2 Equipment/Reagents/Solutions	168
		5.2.2.1.3 Protocol	168
		5.2.2.2 Antigens	168
		5.2.2.2.1 Principle	168

		5.2.2.2.2	Solutions	169
		5.2.2.2.3	Direct Reaction Protocol	170
		5.2.2.2.4	Indirect Reaction Protocol	170
	5.2.2.3	Particular Case: The Biotin Label		171
		5.2.2.3.1	Principle	171
		5.2.2.3.2	Solutions	172
		5.2.2.3.3	Streptavidin Protocol	172
		5.2.2.3.4	Indirect Reaction Protocol	173
5.2.3	Revelation			173
	5.2.3.1	Colloidal Gold		174
		5.2.3.1.1	Principle	174
		5.2.3.1.2	Solutions	174
		5.2.3.1.3	Direct Reaction Protocol	174
	5.2.3.2	Revelation of the Enzymatic Label		175
		5.2.3.2.1	Principle	175
		5.2.3.2.2	Alkaline Phosphatase	175
		5.2.3.2.3	Peroxidase	178
5.2.4	Amplification by Tyramide			180
	5.2.4.1	Tyramide		181
		5.2.4.1.1	Principle	181
		5.2.4.1.2	Tyramide Conjugates	182
		5.2.4.1.3	Advantages/Disadvantages	182
	5.2.4.2	Direct Visualization of Amplification		183
	5.2.4.3	Other Possibilities for Amplification		185

CHOICE OF METHOD

The choice of revelation method depends on the hybrid label:
- Radioactive isotope
- Antigen
- Fluorochrome

Irrespective of the nature of the label, the revelation step involves two parameters:
- Sensitivity
- Resolution

There are three possible approaches, depending on the label:
- Autoradiography

- Immunohistology

- Fluorescence

↝ *See* Chapter 1.

↝ Detection of radiation
↝ Immunological detection
↝ Detection by direct fluorescence or after amplification by immunocytology
↝ The revelation step is the one that materializes the hybrid. A lot of work has been done on this step, which is common to many other methods of morphological analysis.

↝ Autoradiography involves the use of a photographic emulsion to record the emitted radiation:
- Solid emulsion on film or
- The tissue is coated with liquid emulsion.

↝ Immunohistology involves an antigen–antibody reaction, which is visualized:
- By a stained reaction or
- By fluorescence

↝ A fluorescent probe can be visualized:
- By ultraviolet excitation or
- Indirectly, by an immunohistological reaction

5.1 RADIOACTIVE HYBRIDS

A radioactive hybrid emits radiation which is conventionally recorded by autoradiography (i.e., by a photographic emulsion that visualizes it).

5.1.1 Principle

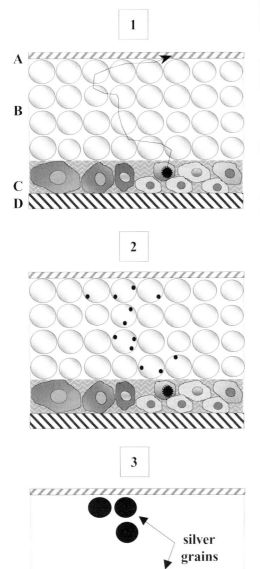

1 = Radiation — The hybrid contains radioactive isotopes, which emit radiation characteristic of the element in question. Autoradiographic detection records these emissions by means of a photographic emulsion.
A = Film support.
B = Photographic emulsion.
C = Tissue or cells.
D = Slide.

2 = Exposure — The emulsion that is placed in contact with the hybrids records the radiation emitted in the course of the exposure, in the form of latent images (•).

3 = Development — These images are turned into visible grains of silver in the course of the photographic development process. Numerous parameters enter into the interpretation of the results.

Figure 5.1 Principle of macroautoradiography.

5.1.2 Characteristics

5.1.2.1 Radiation

The isotopes most commonly used in autoradiography are ^{35}S, ^{33}P, ^3H, and ^{32}P.

The purpose of autoradiography is to visualize radiation or particles emitted by radioactive isotopes through their materialization as grains of metallic silver.

⇝ Only these isotopes, along with β radiation, are recorded by the emulsion.
⇝ See Figure 5.1.

5.1.2.2 Photographic emulsions

These are composed of silver bromide diluted in gelatin. They either take the form of a liquid in a gel (emulsion), or that of a solid (film). They are characterized by:
- The size of the grains
- The thickness of the layer

- The type of medium

⇝ Crucial for resolution
⇝ Modifies the sensitivity and the resolution
⇝ Distinction between macroautoradiography and microautoradiography

5.1.2.3 Exposure

The sample, covered by a film or a thin layer of photographic emulsion, is stored in darkness for a period of time that can vary from a few hours to several months.

⇝ The intensity of the autoradiographic signal depends, first and foremost, on the relationship between the exposure time and the half-life of the isotope. These may be enough to saturate the film or emulsion (with the transformation of all the silver salts into metallic silver).

Exposure time depends on the radioisotope used, the specific activity of the probe, and the abundance of the hybrids (and thus the quantity of sequences looked for in the cell).

⇝ The use of autoradiographic film makes it possible to determine this exposure time without damaging the tissue sections.

5.1.2.4 Development

The revelation process changes the latent images (*see* Figure 5.1(2)) resulting from the activation of the silver salts into visible metallic silver by the action of the radiation.

$$Br^- + \text{radiation} \rightarrow Br + e^-$$
$$Ag^+ + e^- \rightarrow Ag$$

⇝ See Figure 5.1(3).
⇝ There are two chemical steps:
- The first step is that of development, with the changing of the silver salts into metallic silver by the action of radiation,
- The second step is that of fixation, i.e., the dissolution of the remaining salts of silver bromide. There are many methods for reducing silver salts. Fixation, on the other hand, generally uses sodium thiosulfate at 30% as the basic chemical.

5.1.2.5 Efficiency

An efficiency (Figure 5.2) of the order of 15% is generally taken to be acceptable. Such a low level is explained by the fact that not all the latent images are changed into silver grains during the revelation process (15% efficiency means that it takes six disintegrations of the radioisotope to produce 1 grain of silver).

↪ It should be noted that (1) whereas the radiation from the radioactive isotope is emitted in three dimensions, the emulsion is present only on one side of the section; (2) the emulsion is not 100% efficient; (3) the revelation system is not perfect; (4) there is background due to innumerable causes which are inherent in the technique (a control can be carried out by omitting the probe from the hybridization buffer). The efficiency of autoradiography has not been improved for some years now, but it is still an extremely sensitive method.

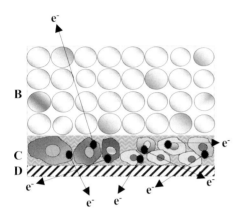

- **B** = emulsion
- **C** = cells
- **D** = slide

Figure 5.2 Efficiency achieved by autoradiography.

5.1.2.6 Localization *in situ*

In *in situ* hybridization, the creation of an image on an autoradiographic emulsion by radiation reveals the position of a nucleic acid, with, for parameters:
- The energy of the radiation
- The size of the grains of emulsion (the precision of the labeling depends on the size of the crystals in the nuclear emulsion)
- The thickness of the layer of emulsion

↪ There always remains a margin of interpretation of the statistics regarding the exact position of the nucleic acid that is being studied, due to disparities between the localization of the silver grains and the source of the emission. The best results are obtained with radioelements whose emitted radiation is not very energetic, in other words, not very penetrating. In this case, the silver grains observed are localized in the neighborhood of the source of the radiation. ^{35}S offers a good compromise between energy and efficiency.

5.1.2.7 Detection

The detection of the signal is possible:
- On the macroscopic scale, on film

↪ With autoradiographic film, the signal can at one and the same time be localized at the macroscopic level and quantified at the level of the tissue.

- On the microscopic scale, in liquid emulsion

↪ With autoradiographic emulsions, localization and quantification can both be carried out at the cell level.

5.1.2.8 Artifacts

Artifactual labeling (background) can have a number of different origins:

- The age of the emulsion
- Irregularities in the section (striations, fissures, etc.)
- Variations in the thickness of the layer of emulsion
- Traces of γ or α radiation
- The presence of chemicals
- Excessive prolongation of the revelation step
- Accidental causes

↝ Not all the grains observed correspond to specific labeling, which therefore has to be quantified and subjected to a statistical analysis.
↝ Can give rise to a shadow on the section.
↝ Can cause variations of emulsion thickness at a certain point, and thus an accumulation of grains.
↝ Can occur during the dipping or drying step, and can also result from accumulations of grains
↝ External radiation
↝ Chemography
↝ The revelation of latent pseudoimages

↝ For example, inappropriate light

5.1.2.9 Quantification

The quantification of radioactivity can be done:
- By macroautoradiographic measurement of the optical density of the autoradiogram
- By microautoradiographic measurement of the number of grains per unit area

↝ Tissue level, rapid method, standardized

↝ Cell level, difficult method

5.1.3 Macroautoradiography

5.1.3.1 Principle

Macroautoradiography is used to visualize, in an overall, homogeneous way, a hybrid within the totality of a section, a tissue, or an organ, by apposition of a film.

↝ Given the fact that it is industrially manufactured, film provides all the advantages of homogeneity and reproducibility, and thus the possibility of making comparisons between signals.

5.1.3.2 Choice of autoradiographic film

There exist different types of autoradiographic film, each with its own particular characteristics.

1. Single-coated film
- Size of grains less than size of grains in double-coated film
- Thickness of the film less than that of double-coated film
- High degree of sensitivity

- Cost generally high

↝ The choice of film depends on the isotope used, the density of the hybrid, and the sensitivity desired.

↝ Good resolution
↝ Diminution of the dispersion of the radiation
↝ More difficult to manipulate

↝ Quite a long exposure time: indicated for use with ^{3}H (not very penetrating β$^-$ radiation)

- Quantification possible

2. Double-coated film
- Higher sensitivity
- Average degree of resolution
- Lower cost

❑ *Advantages*
- Quick to set up
- Short exposure time
- Evaluation of the background, and of the specific signal
- Calculation of exposure time
- Comparison of several sections
- Overall view of an organ or organism
- Selection of slides
- Quantification

❑ *Disadvantages*
- Sensitivity
- Average degree of resolution
- Difficult manipulation
- Artifacts

↪ Radioactivity quantifiable by measuring the optical density of the autoradiograph

↪ Shorter exposure time than for single-coated film
↪ Larger emission cone
↪ Which means that it can be used as a control for exposure time

↪ Rapid response to the test of evaluation of the signal (visualization of a possible signal)
↪ Signal / background ratio

↪ As a function of the background noise
↪ Relative quantitative analysis
↪ Large recording area
↪ For detection on the microscopic scale (liquid emulsion)
↪ Measurement of the homogeneous signal density, using single-coated film

↪ Technique not applicable if the sequences being studied are not very strongly expressed
↪ Macroscopic localization.
↪ Risk of scratching
↪ *See* Section 5.1.2.8

5.1.3.3 Summary of the different steps

↪ *Safelight*

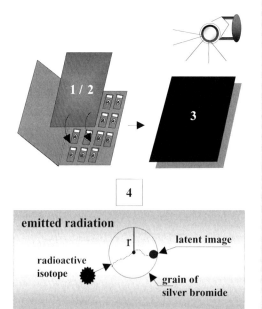

1 = Placing the sections in the cassette.

2 = Positioning the film.

3 = Storage (room temperature).

4 = Macroautoradiographic exposure time.

5.1 Radioactive Hybrids

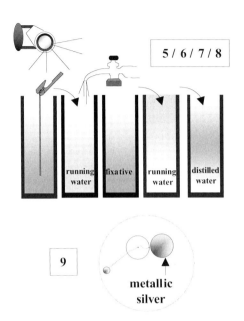

5 = Development.

6 = Rinsing in running water.

7 = Fixation

8 = Rinsing:
- Running water
- Distilled water

9 = Evaluation of the signal

Figure 5.3 Macroautoradiography protocol.

5.1.3.4 Equipment/reagents/solutions

1. Equipment
- Cassette the right size for the film
- Tray the right size for the film

2. Reagents
- Autoradiographic films
  — Single-coated film
  — Double-coated film
- Developer
  — Kodak LX 24
  — D19
  — Universol
- Fixative
  — Kodak 3000
  — Ilford Hypam
  — Sodium thiosulfate

3. Solutions
- Developer

↪ Regular checks should be carried out to make sure that the box is light-tight.
↪ No sterility precautions are necessary.

↪ The size of the film depends on the number of slides to be exposed.
↪ Hyperfilm ^{3}H, Bio Max MR
↪ X AR 500, etc.
↪ The film/developer characteristics (as specified by the manufacturer of the film) must be respected if the signal is to be optimized.

↪ This is the basis of all fixatives.

↪ *See* Appendix B5.1. To be diluted according to the manufacturer's instructions.

- Fixative
  — 30% sodium thiosulfate

↝ *See* Appendix B5.2. To be diluted according to the manufacturer's instructions.
↝ The fixative appears less important than the developer.

5.1.3.5 Protocol

1. Place the fully dry sections in the autoradiographic box. Attach them to the bottom of the box.
2. Put the autoradiographic film in place.
3. Exposure time:
   - If the film is double coated **24 to 48 h**
   - If the film is single coated **Several days**

↝ **All the following steps are carried out in a darkroom, with a safelight.**
↝ Any trace of humidity will cause an autoradiographic chemoreaction, and thus background noise.

↝ Double-coated film can serve as a test for the exposure time of sensitive single-coated film.
↝ Very sensitive film. This type of film is used for quantification purposes. Care should be taken to orient the film correctly: when the rounded corner is at the upper left, the emulsion is on the lower surface, in contact with the slide.

4. Store the boxes at room temperature. Wrap them in aluminum foil, or place them in a bag.
5. Expose

↝ Exposure time depends on the type of double-coated film.

↝ As an example, if the isotope is ^{35}S:
  - For double-coated film, 1 day
  - For single-coated film, 3 to 4 days

These times should be divided by 2 if the isotope is ^{33}P, and multiplied by 3 if it is ^{3}H.

6. Revelation:
   - D 19 **4 min at 17°C**
   - Rinse in running water **1 min**
7. Fixation
   - Fixative **5 min**

↝ Previously cooled

↝ Maximum time: the sections can suffer damage (1 min minimum).

8. Rinse
   - Running water **30 min**
   - Distilled water **1 min**
9. Dry. **at RT, or 37°C**

10. Evaluate the signal.

11. Quantify the signal.

↝ Eliminate possible traces of lime.
↝ Temperature < 50°C, so as not to damage the film.
↝ Define the microautoradiographic exposure time in terms of the intensity of the signal. The presence of a standard range is often useful in determining the state of the film.
↝ Densitometry

5.1.4 Microautoradiography
5.1.4.1 Principle

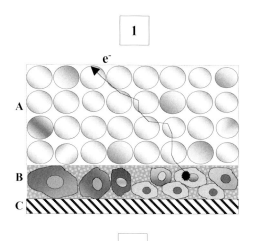

1 = Photographic emulsion — Microautoradiography is used to visualize the silver grains individually, and to match them to cellular structures. The photographic emulsion is applied in liquid form in direct contact with the cells.
- **A** = photographic emulsion
- **B** = cells
- **C** = slide

2 = Exposure — Recording in the form of latent images of radiation emitted during exposure.

3 = Revelation — Transformation of the latent images into visible grains of metallic silver in the course of the photographic development process. Numerous parameters enter into the interpretation of the results.

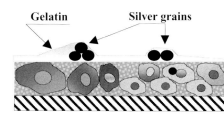

Figure 5.4 Principle of microautoradiography.

5.1.4.2 Choice of autoradiographic emulsion
Different nuclear emulsions can be used, according to the sensitivity desired. Size of the AgBr grains:

• Ilford K5 emulsion	0.2 µm	⇀ High sensitivity
• LM1-Amersham	0.25 µm	⇀ Average sensitivity
• Kodak NTB2 emulsion	0.26 µm	⇀ Average sensitivity

⇀ It is a function of the size of the silver bromide grains in the chosen emulsion.

❑ *Advantages*
• Cellular resolution

• Signal/structure matching
• Quantification on a single section

⇝ According to the isotope used, it is possible to define compartments in the cell.
⇝ This is unlike macroautoradiography.
⇝ The thickness of the layer of emulsion must be controlled.
⇝ This is evaluated by counting the silver grains on a section. It is difficult to compare grain density between two sections.

❑ *Disadvantages*
• Thickness of the section difficult to control.
• Thickness of the layer of emulsion

• Contamination proportional to the number of slides dipped
• Difficult repeatability

• Quantification difficult from one section to another

⇝ Without any particular precaution being taken, and depending on the state of the section, the surface of the emulsion is generally smooth.
⇝ This contamination also varies according to the isotope used ($^{32}P > ^{33}P > ^{35}S > ^{3}H$).
⇝ There are variations of the thickness of the layer of emulsion from one section to another.
⇝ Quantification is difficult due essentially to the thickness and homogeneity of the layer of emulsion. Only relative percentages can be compared.

5.1.4.3 Summary of the different steps

⇝ *Safelight*

1 = **Liquefaction of the emulsion** (43°C)

2 = **Dipping the sections**

3 = **Drying** (2 h to overnight)

4 = **Storage** (4°C)

5 = **Exposure time**

5.1 Radioactive Hybrids

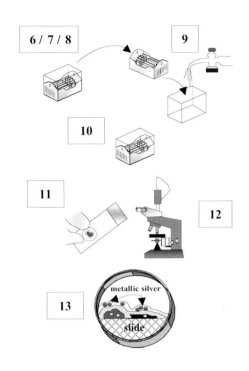

6 = **Development (17°C)**
7 = **Rinsing**
8 = **Fixation**
9 = **Rinsing**
  • Running water
  • Distilled water
10 = **Dehydration** — alcohol in increasing concentrations

11 = **Mounting** (xylene resin)
12 = **Observation**

13 = **Interpretation** — evaluation of the signal

Figure 5.5 Microautoradiography protocol.

5.1.4.4 Equipment/reagents/solutions
1. Equipment — Darkroom

- Autoradiographic boxes
- Beaker / Joly tube / emulsion tube
- Black tape
- Clean histological slides
- Filter paper
- Safelight
- Slide-holder

- Staining trays
- Water bath with the thermostat set to 43°C

2. Reagents
- Alcohol 100%
- Desiccant
- Developer
- Fixative –• Sodium thiosulfate
- Nonaqueous mounting medium
- Nuclear emulsion

↪ All the following steps are carried out in a darkroom, with a safelight placed 1.50 m above the working surface, the humidity in the room being ≈ 20 to 40%.
↪ Clean and light-tight
↪ To dilute the emulsion
↪ Light-tightness of the storage boxes
↪ Bubble test, storage of the desiccant
↪ For wiping the backs of the slides
↪ Autoradiographic room
↪ Always the same one, so that the slides are always at the same angle
↪ Nonsterile
↪ To liquefy the autoradiographic emulsion

↪ Ordinary
↪ Silica gel wrapped in filter paper
↪ Instructions for the use of the emulsion
↪ Or sodium hyposulfite
↪ *See* Appendix B8.2.
↪ *See* Chapter 5.1.4.2.

- Sodium borate ⇝ Neutralization of toluidine blue
- Stains
  — Toluidine blue
  — Methyl-green
- Xylene ⇝ Or another solvent of the mounting resin

3. **Solutions**
   - D 19 developer solution ⇝ *See* Appendix B5.1
   - Distilled water ⇝ Nonsterile
   - 30% sodium thiosulfate ⇝ *See* Appendix B5.2
   - Stains
     — Methyl-green, sodium acetate buffer; acetic acid, pH 4.2 ⇝ *See* Appendix B7.6
     — 0.02% toluidine blue in water ⇝ *See* Appendix B7.1 (stock solution); can be diluted in oxalic acid neutralized by ammonia

5.1.4.5 Coating the slides

After the hybridization and washing steps, the slides are dehydrated, dried, then dipped in the nuclear emulsion. ⇝ *See* Chapter 4.

1. ***Dilution of the emulsion***

   a. Take a beaker, or a specially designed receptacle. Make two marks: one to correspond to the volume of water necessary for the dilution, and a second to correspond to the total volume (water + emulsion).

   ⇝ For 20 to 30 sections, use a Joly tube (volume ≈ 30 mL).
   ⇝ The marks must be visible under a red safelight.

   - LM1-Amersham   **2:1**   ⇝ Emulsion – sterile distilled water 2:1 ($^v/_v$)
   - Kodak NTB2   **1:1**   ⇝ Emulsion – sterile distilled water 1:1 ($^v/_v$)
   - Ilford K5   **1:1**   ⇝ Emulsion – sterile distilled water 1:1 ($^v/_v$)

   b. Fill with water up to the lower mark.
   ⇝ The temperature-regulated water bath should be equipped with beaker-racks.

   c. Leave the beaker and the emulsion at 43°C for 5 to 10 min before dilution.
   ⇝ This will liquefy the emulsion.

   d. Add the emulsion with a porcelain spoon, or else pour it, until the upper mark is reached.
   ⇝ Use the auxiliary safelight placed on the draining board. The lamp should be turned toward the wall when the emulsion is in the liquid phase.
   ⇝ It is possible to dilute all the emulsion, and to store it in this form in darkness, at 4°C.

   e. Mix carefully, and shake every 20 min.
   ⇝ Mix with a glass rod or strip, **slowly** (to avoid producing bubbles); do not use metal instruments.

2. ***Liquefaction***   **1 h at 43°C**
   ⇝ The emulsion must be perfectly homogeneous.
   ⇝ Check the temperature of the emulsion, which must be constant at **43°C**. Too high a temperature denatures the emulsion and causes background.

5.1 Radioactive Hybrids

3. Coating

a. Dip and remove 10 or so clean slides to clean the surface of the emulsion and eliminate the air bubbles.

b. Slowly dip the slides vertically in the emulsion. Remove them with a slow, uniform movement, and drain the excess emulsion on the rim of the beaker for a few seconds, then on a filter paper.

4. Drying — Dry in a vertical position (Figure 5.4) **2 h to overnight at RT**

↪ All the bubbles should disappear.

↪ Avoid ripples, since the thickness of the film must be uniform.
↪ The slide should be dipped until the level of the emulsion is 0.5 cm above the last section (problem of thickness of the layer of emulsion).
↪ Leave one slide without a section (background noise control).
↪ The vertical position ensures the homogeneity of the layer of emulsion.
↪ The slides are dried in air and placed in a slide holder at an angle of 80°, with the sections turned toward themselves.
↪ Wipe the lower part of the slides.
↪ Turn the slides over after 15 to 30 min (optional) to dry the zone of contact between the slide and the rack.

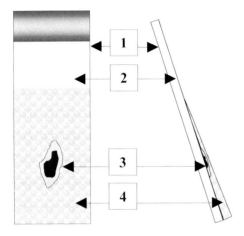

1 = Glass slide
2 = Film holder
3 = Section
4 = Emulsion

Figure 5.6 Drying the emulsion-coated slides.

↪ If the tissue is damp, the dampness will act on the crystals of silver bromide (AgBr), which will result in background.
↪ Use a dehydrating agent of the silica gel type to absorb dampness.
↪ Store at 4°C to limit the development of bacteria in the emulsion.
↪ Plan for several sets of slides to be developed together after different exposure times.

Note: The emulsion must be dry on the slide before storage in the autoradiographic box.

5. Storage — Store the slides in hermetic boxes containing a dehydrating agent wrapped in aluminum foil, or placed in a black bag, at **4°C**.

6. Exposure — Exposure time is around five times longer than for film.

5.1.4.6 Microautoradiographic revelation

Exposure time is difficult to calculate with precision, so revelation should be done at intervals of several days, e.g., the first after 7 days, and the others at further intervals of 7 days.

↪ **The revelation step is carried out in conditions identical to those for the coating of the slides, with a safelight.**
↪ The test of evaluation of radioactivity before the coating of the slides allows the exposure time to be predetermined.

- Take the box out at least 2 h before opening it at the temperature of the laboratory.
- Place the slides in a staining tray and transfer them successively to the following baths.

1. Revelation
- Developer (D19)   **4 min at 17°C**

⇝ To avoid the problem of condensation, the box must be dry before it is opened.

⇝ Possibility of increasing the temperature if the labeling is weak
⇝ Possibility of lowering the temperature if there is a lot of background, which causes a diminution of the signal
⇝ Possibility of diluting the developer to reduce the intensity of the signal

- Rinsing in distilled water   **30 s**

⇝ Rapid

2. Fixation
- 30% sodium thiosulfate   **5 min**

⇝ With delicate tissue, the time can be reduced to 1 min

- Rinsing in running water   **30 min**

⇝ **Note:** Do the rinsing at the same temperature (17 to 18°C) as the revelation. Avoid temperature differences, which could cause folds in the sections.
⇝ Abundantly

- Distilled water   **1 min**

Following steps:
- The preparations are observed without staining:
  — Dark field

⇝ *See* Section 6.3.

- The preparations are stained before observation:
  — Bright field
  — Epipolarization

⇝ *See* Section 6.1.

5.2 ANTIGENIC HYBRIDS

An antigenic hybrid (*see* Chapter 4) has a site that can form an antigen–antibody complex during an immunohistological reaction. Some of them, however, such as the fluorochromes, can be visualized directly by fluorescence.

⇝ The most commonly used haptens are
- Biotin
- Digoxigenin
- Fluorescein (*see* Chapter 1)

⇝ Fluorescein, rhodamine, Rhodol green, etc.

5.2 Antigenic Hybrids

5.2.1 Immunohistological Reaction

The purpose of the immunohistological reaction is to form a complex of high affinity between the antigen and a molecule that is specific to it (tool) for visualizing the antigen, and thus the hybrid.

5.2.1.1 Immunocytological tools

These tools are molecules that have a high degree of affinity with haptens.

↪ These molecules are produced during immunological reactions (antibodies), or spontaneously form a complex with haptens.

5.2.1.1.1 ANTIBODIES

- Immunoglobulin of polyclonal or monoclonal origin
- Fragments of IgG: Fab and F(ab')$_2$

↪ The G immunoglobulins (IgG) are those that are most widely used in immunohistology.

↪ The F(ab')$_2$ immunoglobulin fragment is obtained by enzymatic digestion (cleavage by pepsin) (variable sequence of MW = 100 kDa), which, after reduction, can give rise to 2 Fab' fragments with properties analogous to those of the Fab fragment (Figure 5.7).
↪ The Fab immunoglobulin fragments obtained by enzymatic digestion (action of papain) (variable sequence).
↪ The Fc immunoglobulin fragment is obtained by enzymatic digestion (action of pepsin) (constant sequence).

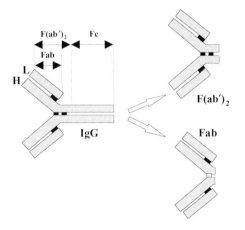

Figure 5.7 Molecule of IgG, and fragments resulting from proteolytic digestion of IgG.

5.2.1.1.2 STREPTAVIDIN
Streptavidin (Figure 5.8) has a very high affinity (Kd = 10^{15} M^{-1}) for biotin. One molecule of streptavidin can bind four biotin molecules by noncovalent binding.

↪ Streptavidin is a homotetramer isolated from *Streptomyces avidinii*.
↪ These advantages, in comparison to avidin, are:
- Nonglycosylated protein
- Neutral isoelectric point
- Without interaction with lectins

Revelation of Hybrids

↪ Similar molecules are available commercially
  • Avidin: a glycoprotein of low molecular mass obtained from egg white, which has 4 fixation sites for biotin
  • Extravidin

Figure 5.8 Streptavidin–biotin complex.

5.2.1.1.3 BIOTIN
One of the properties of this vitamin is its very high-affinity bond with streptavidin (*see* Figure 5.8).

↪ Biotin is used as:
  • A hapten (i.e., labeled nucleotides, *see* Section 1.2.2.1.1);
  • A conjugated label (*see* Section 5.2.1.2.5).

It can be coupled with:
• IgG
• Fab, F(ab')$_2$ fragments
• Streptavidin

↪ In this case it is not conjugated to a label and must be considered an antigen.
↪ Unsaturated complex

5.2.1.2 Labels
Whatever the type of immunohistological reaction, different types of label are used:
• Enzymatic
• Fluorescent
• Particular

↪ These labels are adsorbed onto either immunoglobulin or streptavidin.

5.2.1.2.1 ENZYMES
These are the most widely used labels. The choice of an enzyme depends on the existence of endogenous enzymes in a tissue.

↪ Some enzymes, because they are more stable, are more commonly used.
↪ Numerous chromogens exist for all the different enzymes.
↪ This is a small molecule.

1. Peroxidase — Horseradish peroxidase (black radish extract) is an enzyme of MW = 40 kDa.

❑ *Advantages*
• Numerous chromogens

↪ Precipitates of different colors (*see* Section 5.2.3.2.3)
↪ Complementary to alkaline phosphatase
↪ Possible mounting in resin

• Multiple labeling
• Insoluble in alcohol

❑ *Disadvantages*
• Endogenous enzymes
• Diffusion of the precipitate

↪ Pretreatment of sections by the addition of an inhibitory agent such as 10% hydrogen peroxide, combined or not with sodium nitrite or methanol (*see* Appendix B.6.1.3)

2. Alkaline phosphatase — Alkaline phosphatase (extracted from calf intestine) is a very widely used enzyme of MW = 80 kDa.

❏ *Advantages*
- Numerous chromogenic substrates

- Sensitivity

- Simplicity, reproducibility

- Multiple labeling
- Ease of observation

❏ *Disadvantages*
- Soluble in alcohol

- Cannot be conserved

- Endogenous phosphatase

- Resolution less than that obtained with fluorescent labels

3. β D galactosidase — β D galactosidase (MW = 500 kDa) is not very widely used because it is unstable. It is extracted from *Escherichia coli*.
Its optimal reactive pH is 7.0 to 7.5.

❏ *Advantage*
- No β D galactosidase in biological tissues

❏ *Disadvantage*
- Low sensitivity

5.2.1.2.2 FLUORESCENT LABELS
A fluorochrome is a molecule whose structure includes conjugated double bonds (Figure 5.9) and which, in response to excitation by a photon, emit another photon of a longer wavelength than that of the exciting photon.

↪ This is a large molecule.
↪ This enzyme is very frequently found in biological tissues. Its endogenous activity is inhibited by levamisole or by heat treatment (*see* Appendix B.6.1.2).

↪ Precipitates are of different colors (*see* Section 5.2.3.2.2).
↪ The NBT–BCIP system produces two precipitates with a single alkaline phosphatase molecule.
↪ Ready-to-use chromogenic systems are available.
↪ The conjugates are available in all forms (IgG, Fab, streptavidin, etc.).
↪ There is complementarity to peroxidase.
↪ Use a microscope with a bright field.

↪ This avoids dehydration.
↪ Mounting after dehydration is possible if the reaction is very intense (loss of signal, but attenuation of the background).
↪ Chromogens that are unstable over time are present. Conservation is possible at 4°C.
↪ Pretreatment is indispensable to the inhibition of the signal (*see* Appendix B6.1.2).
↪ There is diffusion of colored precipitates around enzymatic sites.

↪ This is a large molecule.

↪ The 5-bromo-4-chloro-3-indoxyl-β D- galactoside ($C_{14}H_{15}BrClNO_6$) substrate produces a turquoise-blue precipitate.

↪ This avoids the need for an inhibition step.

↪ This structural property is responsible for the fluorescence resulting from excitation by light.

—C=C—C=C—

Figure 5.9 Conjugated double bonds characteristic of fluorochromes.

Each fluorochrome has its own characteristic excitation and emission spectra.

↝ *See* Tables 5.1 and 6.1. Filters can be used to separate the emission spectra of different fluorochromes (wavelength between 300 and 600 nm).

Different types of fluorochrome can be used:
- Rhodol green
- Fluorescein isothiocyanate
- Tetramethyl rhodamine isothiocyanate
- Texas red sulfonyl chloride
- Aminomethylcoumarin

↝ These are classified by effectiveness.

↝ FITC
↝ TRITC

↝ AMCA

❑ *Advantages*
- Rapid method

- Ease of detection
- Resolution

- Multiple labels

↝ No reaction is necessary for visualizing them.

↝ The best resolution is obtained with short wavelengths.
↝ A specific filter is needed for each fluorochrome.

❑ *Disadvantages*
- Autofluorescence
- Absorption of light
- Photodegradation of fluorochromes

↝ Control

↝ Relative to exposure time

1. Rhodol green (Figure 5.10)

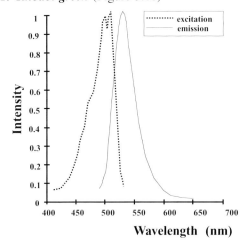

↝ Currently the most effective fluorochrome. It gives an excellent degree of efficiency (sensitivity three times greater than that of FITC for a given excitation).
↝ Useful with small probes
↝ Its half-life is five times greater than that of FITC.
↝ Maximal absorption is at 500 nm, and maximal emission at 525 nm.

Figure 5.10 Absorption and emission spectra for Rhodol green.

2. *Fluorescein isothiocyanate (FITC)*
   (Figure 5.11)

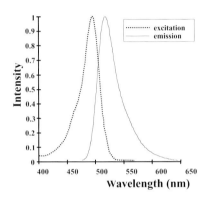

↪ FITC is the fluorochrome that has been most widely used. It emits a yellow-green fluorescence after excitation with blue light.
↪ Maximal absorption is at 490 nm, and maximal emission at 525 nm.

Figure 5.11 Formula (top) and absorption and emission spectra (bottom) of FITC.

3. *Tetramethyl rhodamine isothiocyanate (Figure 5.12)*

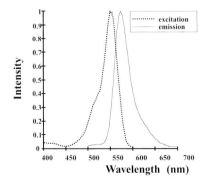

↪ Rhodamine or TRITC

↪ Rhodamine emits a red fluorescence when excited with green light.
↪ Maximal absorption is at 595 nm, and maximal emission at 620 nm (rhodamine filter).

Figure 5.12 Formula (top) and absorption and emission spectra (bottom) of rhodamine.

4. *Texas red sulfonyl chloride (Figure 5.13)*

⇝ Texas red emits a red fluorescence when excited with green light.

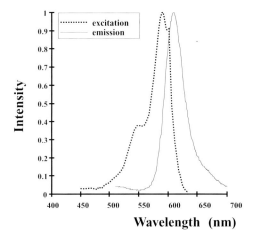

⇝ Maximal absorption is at 595 nm, and maximal emission at 620 nm (rhodamine filter).

Figure 5.13 Formula (top) and absorption and emission spectra (bottom) of Texas red.

5. Aminomethylcoumarin (Figure 5.14)

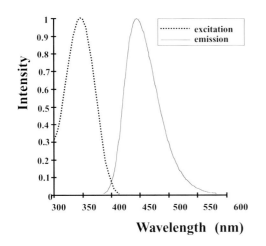

Figure 5.14 Absorption and emission spectra of AMCA.

↝ AMCA

↝ AMCA (7-amino-4-methylcoumarin 3-acetic acid) emits blue fluorescence when excited with UV.

↝ Maximal absorption is at 340 nm, and maximal emission at 425 nm.

Table 5.1 Summary of the Absorption and Emission Wavelengths of Fluorochromes

Fluorochromes	λ Excitation, nm	λ Emission, nm	Stain
AMCA	340	425	Blue
Cyanin	550	580	Red
FITC	490	525	Yellow-green
Oregon green 500	503	522	Green
Rhodamine	540 – 560	570	(Orange) (red)
Texas red	595	620	Orange
Rhodol green	500	525	Green

5.2.1.2.3 PARTICLE LABEL

Colloidal gold is a very widely used label, both in light and in electron microscopy.

There exists a large variety of gold particles whose size varies from 1 to 40 nm.

The sensitivity of detection decreases with the size of the particle. To increase the detection signal, it is preferable to use gold particles of 1 nm, which are latensified by deposition of silver salts.

↝ This particle is opaque to electrons.

↝ Resolution in light microscopy does not allow them to be visualized.

↝ Silver salts precipitate on contact with colloidal gold particles (see Figure 5.24).

Revelation of Hybrids

❏ *Advantages*
- Visualization by bright field and epipolarization
- Quantification

⇌ *See* Section 6.3.4

⇌ The signal is evaluated according to the number of gold particles / unit area.
⇌ Criteria are the same as for autoradiography (*see* Section 5.1.2.9)

- Double labeling possible

⇌ Double labeling is possible in association with an enzymatic immunohistological reaction.

❏ *Disadvantages*
- Difficulty with reproducibility

⇌ The latensification reaction does not produce particles of identical size.
⇌ Standardization of the latensification process is difficult (*see* Section 5.2.3.1).

- Specificity of latensification

⇌ Silver salts can be deposited on all the particles.

- Large numbers are impossible
- Particles of different sizes after latensification

⇌ Latensification time is crucial.
⇌ It is difficult to distinguish between two particles attached to just one.

- Size

⇌ Because the size of the colloidal gold particles is below the resolving power of optical microscopes, these particles are latensified by the use of silver salts.

5.2.1.2.4 CHOICE OF LABEL
This depends on various parameters, such as:
- The molecule to be labeled

⇌ All the labels for all the different tools are commercially available.

- The presence of endogenous enzyme in the cells or tissues studied

⇌ Most endogenous enzymatic activities can be inhibited, but this increases the duration and complexity of the reaction.

- The thickness of the section

⇌ The penetration of a label depends on its molecular mass.

- The required sensitivity

⇌ The intensity of the nonradioactive signal depends largely on the effectiveness of the detection and amplification systems.

- The conservation of the reactions

⇌ Fluorescent labels cannot be conserved for very long.
⇌ Reactions that use alkaline phosphatase are conserved in an aqueous mounting medium.

- Quantification
- Multiple reactions
- Amplification

⇌ To estimate gene expression
⇌ Complementary labels
⇌ The density of the hybrids to be detected.

5.2 Antigenic Hybrids

Table 5.2 Recapitulative Table of the Characteristics of the Different Labels

	Labels		
	Enzymatic	**Fluorescent**	**Particle**
Sensitivity	Good	Presence of a threshold	Poor
Multiple detections	Enzyme/enzyme Enzyme/particle Enzyme/fluorochrome	Fluorochrome/fluorochrome Fluorochrome/enzyme	Particle/enzyme
Quantification	Difficult	Integration of the emitted fluorescence	Counting
Amplification	Possible	Detection threshold	Indispensable
Background	Presence of endogenous enzyme	Endogenous fluorescence	Due to amplification
Observation	Bright field	Fluorescence	Bright field; epi-polarization
Conservation	Short/long	Short	Long

5.2.1.2.5 PRINCIPLE OF OBTAINING CONJUGATES

1. Conjugates – antibodies (Figure 5.15)

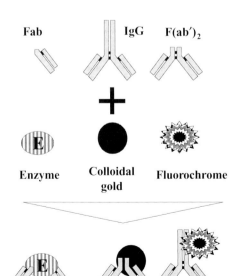

↪ Given that labeling is difficult to reproduce, it is preferable to use commercially available products.
↪ Immunoglobulin, or its fragments, can be conjugated with all the labels previously mentioned.
↪ The experimental conditions for the preparation of labels and their adsorption necessitate specific conditions for each label and each tool.
↪ According to its size, a label can form a complex, either with one or several immunoglobulins, or with their fragments.

Figure 5.15 Principle of obtaining immunoglobulins and their conjugated Fab fragments.

2. Streptavidin conjugates (Figure 5.16)

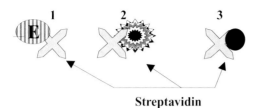

↪ The same conjugates are obtained with streptavidin.

1 = Enzyme
2 = Fluorochrome
3 = Colloidal gold

Figure 5.16 Streptavidin conjugates.

3. Biotin conjugates (Figure 5.17)

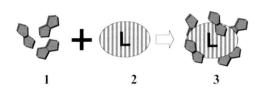

↪ The low molecular mass of biotin means that several molecules attach to each molecule of the label (all the biotin conjugates are commercially available).

1 = Biotins
2 = Label
3 = Conjugate

Figure 5.17 Biotin conjugates.

4. Biotin – streptavidin conjugates (Figure 5.18)

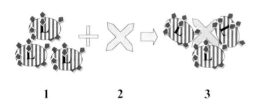

↪ The concentration of conjugated biotin must not be saturating for the binding sites of streptavidin with biotin (*see* Section 5.2.1.1).

1 = Conjugated biotins
2 = Streptavidin (*see* Figure 5.8)
3 = Conjugated complex

Figure 5.18 Biotin–streptavidin conjugates.

5. Peroxidase – anti-peroxidase conjugates (Figure 5.19)

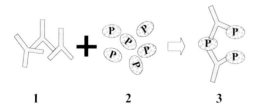

↪ The IgG must be from the same species as the primary antibody used to detect the hapten.
↪ The respective concentrations of IgG and peroxidase are crucial for the saturation of all the antigenic sites.

1 = IgG anti-peroxidase
2 = Peroxidase
3 = Peroxidase–anti-peroxidase complex

Figure 5.19 Formation of the peroxidase–anti-peroxidase complex.

5.2.2 Detection

The revelation of antigenic hybrids (probes containing nucleotides labeled by fluorochromes or haptens) is carried out after the hybridization and washing steps by:
- Fluorescence
- An immunohistological reaction

⇝ *See* Chapter 4.

⇝ *See* Section 5.2.2.1.
⇝ Direct (*see* Section 5.2.2.2.3) or indirect (*see* Section 5.2.2.2.4) immunohistological reaction

5.2.2.1 Fluorescence

This is a direct visualization technique (Figure 5.20).

5.2.2.1.1 PRINCIPLE

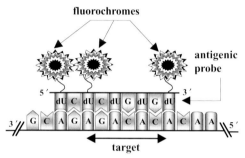

❑ *Advantages*
- Rapid detection

- Possibility of washing controls

- Contrast in the section
- Multiple cell labeling

❑ *Disadvantages*
- Endogenous fluorescence
- The sensitivity of labeled nucleotides to light
- Low sensitivity

- Quality of the labeling

- Short-term conservation of labeled probes
- Conservation of the signal

⇝ The hybrid carries a label that can be detected on its own.
⇝ Only fluorochrome is usable. It is linked to the nucleotide and incorporated into DNA probes (random priming or nick translation), oligonucleotides (3' extension); and RNA probes (transcription *in vitro*) (*see* Chapter 1).

Figure 5.20 Principle of direct visualization of a fluorescent hybrid.

⇝ After rapid observation at the end of the hybridization step, it is possible to determine the efficacy of the different washing steps directly by observing the signal.
⇝ After each washing, it is possible to check the signal/background ratio.
⇝ Counterstaining is possible.
⇝ Unlike chromogens, there is no precipitate with a label that can mask the previous one, and both fluorochromes are visible.
⇝ Possibility of revealing two or more sequences in the same cell, with the availability of numerous sets of filters.

⇝ No inhibition is possible.
⇝ Storage is difficult.
⇝ The density of the label must be sufficient to ensure detection.
⇝ It is very difficult to incorporate a large quantity of fluorochrome-bearing nucleotides without modifying the properties of the probe.
⇝ Fluorochromes are difficult to conserve.
⇝ Duration is always short.

5.2.2.1.2 EQUIPMENT/REAGENTS/SOLUTIONS

- Slides
  ↬ Sterilized (for preparation, *see* Appendix A1.1)

- Mounting medium
  ↬ *See* Appendix B8.3 (amplification of fluorescence with an antifading medium)

- SSC 20X buffer, pH 7.0
  ↬ *See* Appendix B3.4

5.2.2.1.3 PROTOCOL

↬ Only washing steps are required.
↬ Fluorescence varies according to the pH of the preparation (in the case of fluorescein, the fluorescence is maximal at pH 8.0).

1. Check the signal/background ratio, and, if necessary, rinse again:
   - SSC 1X   **5 to 15 min**
   - SSC 0.5X   **5 min**
2. Mount between the slide and the coverslip

↬ Check with a fluorescence microscope.

↬ Check the signal/background ratio.
↬ An antifading mounting is recommended (*see* Appendix B8.3).

3. Observe
   - Directly
   - Using a low-energy camera

↬ *See* Chapter 6.
↬ Use high detection threshold.
↬ This lowers the detection threshold.

5.2.2.2 Antigens

5.2.2.2.1 PRINCIPLE

For this form of visualization, there are two types of reaction:

↬ Whatever the method used, sensitivity and resolution can be modified in the course of this reaction.

1. Direct reaction (Figure 5.21)

↬ Hybrids containing one or more haptens can be detected by a primary antibody (IgG), or fragments conjugated to a label.
↬ Utilization of a single conjugated molecule (*see* Section 5.2.1.2.5)

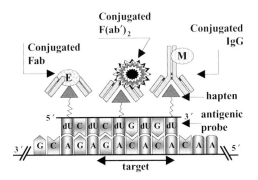

Figure 5.21 Principle of the direct immunohistological reaction.

❑ *Advantages*
- Rapid
- Sensitivity greater than direct visualization
- Amplification possible

❑ *Disadvantage*
- Low sensitivity

↬ A single step

↬ Remains limited

↬ Essentially due to steric hindrance, and to the low level of antigenic labeling of the probes

2. Indirect reaction (Figure 5.22)

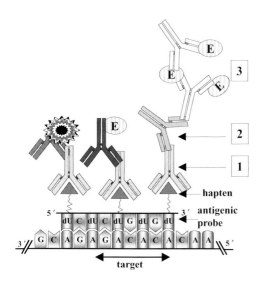

↪ With the indirect reaction, the first steps involve the formation of a complex with haptens by a succession of antigen–antibody reactions. The last step involves a conjugated molecule (*see* Section 5.2.1.2.5, and Figure 5.19).

1 = **Anti-hapten** (species X) (IgG)
2 = **Anti-species X:**
 • Conjugated to a label (IgG, Fab, F(ab')$_2$)
 • Nonconjugated (IgG)
3 = **System of amplification**

Figure 5.22 Principle of the indirect immunohistological reaction.

❏ *Advantages*
• Sensitivity greater than the direct method

• Amplification of the signal

❏ *Disadvantage*
• Long reaction time

↪ Several secondary antibodies can fix to the primary antibody.
↪ There is a cascade of antigen–antibody reactions.

↪ There are successive incubations with each component.

5.2.2.2.2 SOLUTIONS
1. Antibodies
   • Nonconjugated anti-hapten IgG

   • IgG, F(ab')$_2$, Fab:
   — Conjugated anti-hapten
   — Anti-species X conjugate
2. Inhibition of endogenous enzymatic activities
   • Phosphatases
   • Peroxidases
3. Buffers
   • Blocking buffers:
   — 50 mM Tris–HCl buffer; 300 mM NaCl; 1% serum albumin
   — 50 mM Tris–HCl buffer; 300 mM NaCl; 2% goat serum; 0.1 to 0.01% Triton X-100
   • 50 mM Tris–HCl buffer, pH 7.6

↪ Indirect reaction
↪ Monoclonal or polyclonal IgG
↪ Possible to use any label
↪ Direct reaction
↪ Indirect reaction

↪ *See* Appendix B6.1.2
↪ *See* Appendix B6.1.3

↪ *See* Appendix B6.1
↪ Possible to add other agents to the blocking buffer (*see* Appendix B6.1.1), e.g.:
 • Fish gelatin
 • Ovalbumin
 • Unsweetened skimmed milk
↪ Tris–HCl (*see* Appendix B3.6.1)

Revelation of Hybrids

• 50 mM Tris–HCl buffer; 300 mM NaCl, pH 7.6
4. Triton X-100

↝ Tris–HCl / NaCl (*see* Appendix B3.6.5)
↝ Detergent useful, but not indispensable (*see* Appendix B2.24)

5.2.2.2.3 DIRECT REACTION PROTOCOL

↝ This is a short protocol, 90 to 120 min.
↝ All the different steps carried out at room temperature

1. Rinsing
   • Buffer — Tris–HCl/NaCl   **10 min**

↝ Balances the osmolarity of the tissue after the posthybridization washing steps

2. Blockage of the nonspecific sites
   • Blocking buffer   **15 to 30 min**

↝ **Indispensable** step for eliminating any reaction in nonspecific sites, as the sections are preincubated with a nonspecific serum
↝ Possible for the Triton X-100 concentration to vary from 0.01 to 0.3%, according to the thickness of the section.

3. Elimination of excess buffer

↝ Either by aspiration or by carefully wiping the part of the slides around the tissue with filter paper; possible to lay down a circle of hydrophobic material to limit the dilution of the following compound

4. Inhibition of endogenous enzymes

↝ **Optional** (*see* Appendices B6.1.2 and B6.1.3)

5. Spreading the conjugated antibody
   • Diluted in buffer —   **≥ 20 µL/section**
   Tris–HCl/NaCl

↝ Formation of the hapten–antibody complex
↝ Primary antibody: (IgG), conjugated Fab fragments (*see* Section 5.2.1.1)
↝ Dilution of the antibody always weak to compensate for the low level of sensitivity (between 1:10 and 1:100, depending on the density of the labeling)
↝ Possible for too high a concentration of Triton to cause background
↝ Possible use of a staining tray.

6. Incubation of the antibody   **2 h to overnight**
7. Rinsing
   • Tris–HCl/NaCl buffer   **3 × 10 min**

Following steps:
• Revelation
• Observation

↝ Moisture chamber (water on filter paper).

↝ Signal/noise ratio generally low

↝ *See* Chapter 5.2.3.
↝ *See* Chapter 6.

5.2.2.2.4 INDIRECT REACTION PROTOCOL

↝ All the following steps are carried out at room temperature. It is possible to spread a circle of hydrophobic material, to limit the quantity of solution spread on the section.

1. Rinsing
   • Tris–HCl/NaCl buffer   **10 min**

↝ Balances the osmolarity of the tissue after the posthybridization washing steps (*see* Appendix B3.6.5)

2. Blockage of nonspecific sites
   • Blocking buffer   **15 to 30 min**

   ↪ An **indispensable** step for eliminating any reaction in nonspecific sites; sections preincubated with a nonspecific serum

3. Inhibition of endogenous enzymatic activity

   ↪ **Optional** (*see* Appendix B6.1.2 / B6.1.3)

4. Rinsing
   • Tris–HCl / NaCl buffer   **3 × 10 min**
5. Elimination of excess buffer

   ↪ Risk of dilution during the following step

6. Spreading the antibody
   • Diluted in Tris–HCl / NaCl buffer   **≥ 20 µL/section**

   ↪ Formation of the hapten–antibody complex
   ↪ Dilution between 1:50 and 1:500

7. Incubation of the antibody   **60 to 90 min**

   ↪ Moisture chamber (water on filter paper)

8. Rinsing
   • Tris–HCl/NaCl buffer   **3 × 10 min**

   ↪ Or Tris–HCl buffer

9. Elimination of excess buffer
10. Spreading the conjugated antibody
    • Diluted in Tris–HCl/ NaCl buffer   **≥ 20 µL/section**

    ↪ Detection of the complex
    ↪ Secondary antibody: IgG, Fab-fragment, or biotin conjugates, diluted 1:25 to 1:50 (according to the manufacturer's instructions).
    ↪ Dilution less than that of the first antibody

11. Incubation of the antibody   **60 to 90 min**

    ↪ Moisture chamber (water on filter paper)

12. Rinse
    • Tris–HCl/NaCl buffer   **3 × 10 min**
13. To eliminate excess buffer

    ↪ On sections (≥ 100 µL) or in trays

Following steps:
• Revelation
• Observation

↪ *See* Section 5.2.3.
↪ *See* Chapter 6.

5.2.2.3 Particular case: the biotin label
5.2.2.3.1 PRINCIPLE
See Figure 5.23.

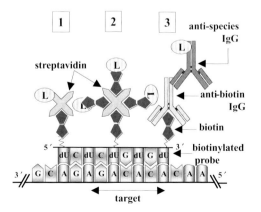

1 = **Direct reaction** — Streptavidin is conjugated to a label.
2 = **Indirect reaction** — The first step uses native streptavidin, and in the second the sites that are free of streptavidin are saturated with conjugated biotin. Conjugated streptavidin – biotin complexes are commercially available.
3 = **Direct or indirect immunohistological reaction** — The reaction uses an anti-biotin IgG. This property is used to amplify the signal in the indirect reaction.

Figure 5.23 Principle of the detection of a biotinylized hybrid.

❏ *Advantages*
- Sensitivity

- Specificity
- Numerous labels available

❏ *Disadvantage*
- Endogenous biotin

⇝ Due to the high affinity of streptavidin for biotin

⇝ Multiple labelings

⇝ Necessary to endogenous biotin, which is present in some animal tissues (kidney, heart, muscle, liver...)

5.2.2.3.2 SOLUTIONS

1. Antibodies
   - Anti-biotin IgG

   - Anti-species conjugated IgG

   - Goat serum

   - Streptavidin

⇝ Conjugated (direct immunohistological reaction) or nonconjugated (indirect immunohistological reaction)
⇝ *See* Figure 5.23,3 indirect immunohistological reaction
⇝ Nonspecific antibody, which can be replaced by a blocking solution
⇝ Conjugated (*see* Figure 5.23,1) or not (*see* Figure 5.23,2)

2. Conjugated streptavidin – biotin complex

⇝ Possible to produce this complex in two steps on the section; (*see* Figure 5.23,2):
   - Nonconjugated streptavidin
   - Conjugated biotin

3. Inhibition of endogenous enzymatic activities
   - Phosphatases
   - Peroxidases

⇝ *See* Appendix B6.1.2
⇝ *See* Appendix B6.1.3

4. Buffers
   - Blocking buffers
     — 50 mM Tris–HCl buffer; 300 mM NaCl; 1% serum albumin
     — 50 mM Tris–HCl buffer; 2% goat serum; 0.1% Triton X-100

   - 50 mM Tris–HCl buffer, pH 7.6
   - 50 mM Tris–HCl buffer; 300 mM NaCl; pH 7.6

⇝ *See* Appendix B3
⇝ *See* Appendix B6.1.1
⇝ The purpose of the high concentration of Na$^+$ ions is to preserve the hybrids.
⇝ It is possible to add other agents to the blocking solution, such as:
   - Fish gelatin
   - Ovalbumin
   - Unsweetened skimmed milk
⇝ Tris–HCl buffer (*see* Appendix B3.6.1)
⇝ Tris–HCl/NaCl buffer (*see* Appendix B3.6.5)

5.2.2.3.3 STREPTAVIDIN PROTOCOL

1. Blockage of nonspecific sites
   - Blocking buffer   **10 min**

2. To form the biotin–streptavidin complex
   - Conjugated streptavidin diluted 1:20 to 1:50 in Tris–HCl/NaCl buffer   **100 µL/section**
     60 to 90 min

   ↪ The dilution depends essentially on the label used.
   ↪ Streptavidin can be replaced by an anti-biotin–IgG conjugate.

3. Rinsing
   - Tris–HCl/NaCl buffer   **20 min**

   ↪ A change of buffer is sometimes advisable.

Following steps:
- Revelation
- Observation

↪ *See* Section 5.2.3.
↪ *See* Chapter 6

5.2.2.3.4 INDIRECT REACTION PROTOCOL
1. Blockage of nonspecific sites
   - Blocking buffer   **10 min**

   ↪ An inhibition step for endogenous biotins may be introduced immediately before incubation with streptavidin.

2. Formation of the biotin–streptavidin complex
   - Streptavidin 1:50 in Tris–HCl/NaCl buffer   **100 µL/section**
     60 to 90 min

   ↪ The dilution depends essentially on the label used.

3. Rinsing
   - Tris–HCl/NaCl buffer   **20 min**

   ↪ This balances the osmolarity of the tissue after the posthybridization washing steps.

4. Incubation of the conjugated biotin
   - Diluted 1:20 to 1:50 in Tris–HCl buffer   **60 min**

   ↪ The dilution depends essentially on the label used.

5. Rinsing
   - Tris–HCl buffer   **20 min**

   ↪ A change of buffer is sometimes necessary for the revelation step.

Following steps:
- Revelation
- Observation

↪ *See* Chapter 5.2.3.
↪ *See* Chapter 6.

5.2.3 Revelation

Except in the case of the direct visualization of fluorochromes (*see* Section 5.2.2.1), labels require a further step to be observed by light microscopy.

5.2.3.1 Colloidal gold

5.2.3.1.1 PRINCIPLE
See Figure 5.24.

conjugated IgG   silver salts   visible product

➯ The density of immunolabeling with a colloidal gold label is inversely proportional to the size of the particles. To obtain a signal, it is necessary to use particles that are smaller than the resolving power of the light microscope. The purpose of latensification is to increase their size.

Figure 5.24 Latensification of colloidal gold.

❏ *Advantages*
- Particular reaction

- Reaction unaffected by alcohol
- Multiple labeling
- Possible counterstaining

❏ *Disadvantages*
- Parasitic reactions

- Necessary to standardize the reaction.

- Reaction carried out with small numbers of slides

➯ Observation in bright field or epipolarization
➯ Possible conservation over time
➯ Carried out last
➯ *See* Table 6.1

➯ Background; certain impurities cause a reduction of silver salts, and thus of nonspecific signal
➯ Temperature, duration, and reagents (controls to be carried out for each reaction)
➯ A control necessary

5.2.3.1.2 SOLUTIONS
- Distilled water
- Amplification solution

➯ Nonsterile
➯ Commercial solution, with expiration date

5.2.3.1.3 DIRECT REACTION PROTOCOL

1. Rinsing
- Distilled water   3 × 1 min

2. Latensification of gold
   a. Incubation in the presence of silver salts solutions, in darkness   **Time to be determined**

   b. Rinsing in distilled water   **1 min**

   c. Counterstaining
   d. Mounting

   e. Observation

➯ All the following steps are carried out at room temperature.
➯ Rinse to reduce aspecific bonds (risk of background in amplification).

➯ The reaction depends on the temperature.
➯ The size of the gold particles is proportional to the incubation time.
➯ Rinse to stop the reaction.

➯ **Optional step** (*see* Appendix B7).
➯ *See* Appendix B8 (any of the different mounting media can be used).
➯ *See* Section 6.2.8.

5.2.3.2 Revelation of the enzymatic label
5.2.3.2.1 PRINCIPLE
See Figure 5.25.

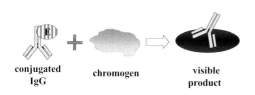

Enzymatic activity is revealed by stained reactions which can use different chromogens according to the desired color of the precipitate.

↪ *See* Chapter 6.4.1.

↪ The activity of the enzyme (E) catalyzes the precipitation of the chromogen as a colored substance which is deposited at the reaction site.

Figure 5.25 Principle of revelation of enzymatic activity.

↪ *See* Appendix B6.2 and the recapitulative table of the different enzymes and their chromogens (*see* Section 6.4 and Table 6.3).

5.2.3.2.2 ALKALINE PHOSPHATASE
The most widely used chromogens (substrates) with alkaline phosphatase are:
- NBT–BCIP
- Fast–Red

1. NBT – BCIP (Figure 5.26)

- NBT, or
Nitroblue tetrazolium
- BCIP or
5-bromo-4-chloro-3-indolyl phosphate

❏ *Advantages*
- Sensitivity

- Stability of the precipitate
- Compatible with multiple labeling

❏ *Disadvantages*
- Reaction sometimes long
- Precipitate soluble in alcohol

↪ These two substrates are soluble in dimethylformamide.
↪ $C_{40}H_{30}Cl_2N_{10}O_6$
↪ MW = 817.70
↪ $C_8H_6NO_4BrCIP \times C_7H_9N$
↪ MW = 433.60

↪ This is due to the formation of two precipitates.
↪ **Note:** soluble in alcohol.

↪ Revelation can take up to several hours.
↪ If the labeling is very intense, rapid dehydration is possible (diminution of the background).
↪

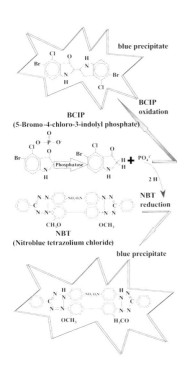

↪ The stained reaction is one of oxido-reduction (oxidation of BCIP and reduction of NBT) which gives precipitates that are insoluble in water but soluble in ethanol.
↪ It gives the highest efficiency (best signal / background ratio).

Figure 5.26 Oxido-reduction reaction of the NBT–BCIP system.

❑ *Reagents / Solutions*
- Dimethylformamide
- Levamisole 100X
- Substrates
  — NBT
  — BCIP
- Tris/NaCl/MgCl$_2$ buffer; pH 9.5

❑ *Protocol*
a. Put the substrate in place:
- NBT–BCIP   **1:250**
                 100 µL/section

b. Incubate under visual surveillance until the desired reaction is obtained.   **1 to 24 h at RT in darkness**

↪ (CH$_3$)$_2$NOCH
↪ C$_{11}$H$_{12}$N$_2$S (*see* Appendix B6.1.2.1)

↪ *See* Appendix B3.6.6.

↪ Preparation (*see* Appendix B6.2.1.1)

↪ **Optional.** Add an enzyme-blocking agent (1 m*M* levamisole, *see* Appendix B6.1.2.1).
↪ It is possible to accelerate the reaction by keeping the slides at 37°C.

c. Stop the reaction in distilled water. **5 min**
d. Counterstain.
e. Mount in an aqueous medium.

↪ If the revelation is insufficient, deposit a further 100 µL of the NBT–BCIP solution.
↪ *See* Section 6.1.3.
↪ The precipitate is soluble in alcohol.
↪ *See* Appendix B8.1.

2. Fast–Red (Figure 5.27)

↪ This is especially suitable for staining smears. The stained reaction is expressed as a substance that is soluble in alcohol. It is necessary to do a counterstain and aqueous mounting. Ready-to-use Fast–Red solutions are commercially available.

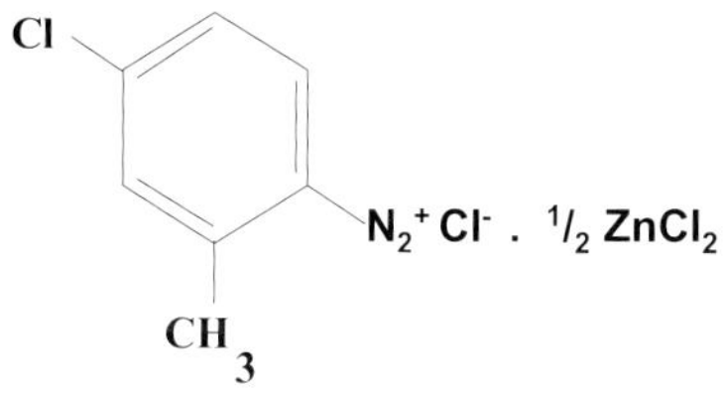

Figure 5.27 Fast–Red formula.

❏ *Advantages*
- Sensitivity
- Stability of the precipitate
- Compatible with multiple labels

↪ **Note:** Soluble in alcohol.

❏ *Disadvantages*
- Reaction sometimes long
- Precipitate soluble in alcohol

↪ Revelation can take up to 24 h.

❏ *Reagents / solutions*
- Dimethylformamide
- Fast–Red

- Levamisole 100 X
- Naphthol phosphate
- Tris/NaCl/MgCl$_2$ buffer, pH 9.5

↪ $(CH_3)_2NOCH$
↪ $C_7H_6N_3O_2$ (5-chloro-2-methoxy-benzene-diazonium chloride)
↪ MW = 250.90
↪ 1 *M* (*see* Appendix B6.1.2.1)
↪ *See* Appendix B3.6.6

❏ *Protocol*
a. Place the filtered substrate directly on the slides:
— Fast–Red **100 µL/section**

↪ Preparation (*see* Appendix B6.2.1.2)

b. Incubate under visual surveillance until the desired reaction is obtained. **1 to 24 h at RT in darkness**
c. Rinse with distilled water. **1 min**
d. Counterstain.
e. Mount in an aqueous medium.

↪ The reaction is finished when the colored precipitate is red and clearly visible. It is possible to accelerate the reaction by keeping the slides at 37°C.
↪ Rinse to stop the reaction.
↪ **Necessary** (*see* Sections 6.1.2 and 6.1.3).
↪ The precipitate is soluble in alcohol.
↪ *See* Appendix B8.1.

5.2.3.2.3 PEROXIDASE

The peroxidase activity (oxidation of the appropriate substrate) produces an insoluble colored substance (precipitate) which materializes the reaction. The chromogens (substrates) appropriate to peroxidase are:
- 3'-Diaminobenzidine tetrachloride (DAB)
- 4-Chloro-1-naphthol
- 3-Amino-9-ethylcarbazole (AEC)

↬ The electron donor is hydrogen peroxide.

1. DAB (Figure 5.28)

↬ The substrate is oxidized in the presence of peroxidase, and produces a signal that is expressed by a yellow-brown precipitate.

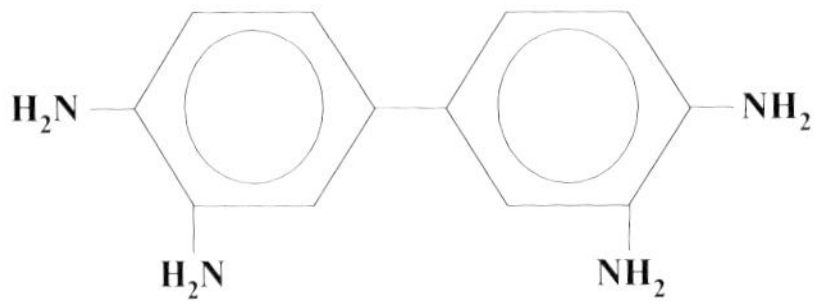

Figure 5.28 DAB formula.

❑ *Advantages*
- Yellow-brown precipitate
- Precipitate insoluble in alcohols

- Counterstaining possible

- Very intense reaction
- Double labeling possible

❑ *Disadvantages*
- Dangerous in powder form
- Possibility of background noise

- Requires hydrogen peroxide of excellent quality
- Dangerous waste

❑ *Reagents/solutions*
- Diaminobenzidine tetrahydrochloride
- 30% hydrogen peroxide
- Tris / NaCl / $MgCl_2$ buffer; pH 7.6

❑ *Protocol*
  a. Put the substrate in place.   **100 µL/section**

↬ Can be intensified by nickel salts
↬ Mounting in resin that is stable over time
↬ *See* Table 6.1
↬ Diminution of the background noise

↬ To be carried out as a second reaction

↬ Can be attenuated by preincubation with DAB solution, without hydrogen peroxide
↬ The stability of hydrogen peroxide limited to a few weeks
↬ Destroy by sodium hypochlorite

↬ $C_{12}H_{14}N_{14} \cdot HCl$ (DAB)

↬ *See* Appendix B3.6.6

↬ For preparation, *see* Appendix B6.2.2.1.
↬ **Optional**. Add an enzyme-blocking agent (endogenous peroxidases, *see* Appendix B6.1.3) before the use of peroxidase conjugates.

b. Incubate under visual surveillance until the desired reaction is obtained. **3 to 10 min at RT**

↪ Observe brown coloring.
↪ A too-long revelation time causes a generalized coloring of the tissue.
↪ Counterstaining is possible.

c. Stop the reaction by dipping the slides in distilled water.
d. Counterstain.

↪ See Section 5.2.6 (hematoxylin is recommended, see Appendix B.4)
↪ The precipitate is insoluble in alcohol.
↪ See Appendix B8.2

e. Mount in resin.

2. 4-Chloro-1-naphthol (Figure 5.29)

↪ The reaction is expressed as a blue-violet precipitate.

Figure 5.29 4-Chloro-1-naphthol formula.

❑ *Advantages*
- Blue precipitate

↪ A weakly different blue from that of the alkaline phosphatase reaction using NBT–BCIP
↪ See Chapter 6 and Table 6.1.

- Double labeling
- Counterstaining possible

❑ *Disadvantages*
- Soluble in alcohols
- Unstable over time

↪ Mounting in an aqueous medium

❑ *Reagents / Solutions*
- Alcohol 95%
- 4-Chloro-1-naphthol
- 30% hydrogen peroxide
- Tris – HCl / NaCl buffer; pH 7.6

↪ $C_{10}H_7ClO$; MW = 178.60

↪ See Appendix B3.6.5

❑ *Protocol*
a. Put the substrate in place. **100 µL / section**

↪ For Preparation, see Appendix B6.2.2.2.
↪ **Optional.** Add an endogenous enzyme-blocking agent (*see* Appendix B6.1.3).

b. Incubate under visual surveillance until the desired reaction is obtained. **5 to 10 min at RT**

↪ The signal is a blue reaction, to be checked by microscope.

c. Stop the reaction with distilled water.

↪ Counterstaining is possible (*see* Section 6.1.3).

d. Mount in an aqueous medium.

↪ The precipitate is soluble in alcohol (*see* Appendix B8.1).

3. 3-Amino-9-ethylcarbazole (AEC) (Figure 5.30)

↪ This substrate is soluble in dimethylformamide.
↪ The reaction is expressed as a red precipitate.

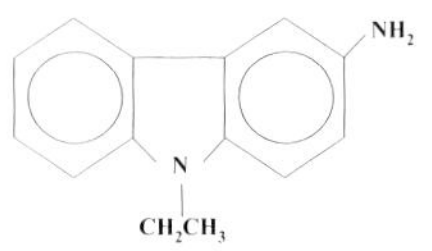

Figure 5.30 AEC formula.

❏ *Advantages*
- Red, clearly visible precipitate
- Double labeling
- Counterstaining possible

❏ *Disadvantage*
- Toxic

↪ The solvent is harmful if inhaled.

❏ *Reagents / Solutions*
- 100 mM acetate buffer; pH 5.2
- AEC (3-amino-9-ethyl carbazole)

↪ $C_{14}H_{14}N_2$
↪ MW = 210.30
↪ $(CH_3)_2NOCH$

- Dimethylformamide
- 30% hydrogen peroxide

❏ *Protocol*

↪ For preparation, *see* Appendix B6.2.2.3.
↪ **Optional.** Add an enzyme-blocking agent (endogenous peroxidases, *see* Appendix B6.1.3).

a. Put the substrate in place. **100 µL / section**

b. Incubate under visual surveillance until the desired reaction is obtained. **10 min in darkness**

↪ Oxidized AEC forms a pink-red precipitate; reaction to be checked by microscope.

c. Stop the reaction with distilled water.

d. Mount in an aqueous medium.

↪ Counterstaining is possible with hematoxylin (*see* Section 6.1.3).
↪ The precipitate is soluble in alcohol (*see* Appendix B8.1).

5.2.4 Amplification by Tyramide

The amplification of the detection of antigenic hybrids is possible, and sometimes necessary. It is carried out after the hybridization and washing steps, during revelation.

↪ This can solve certain problems concerning the sensitivity of the *in situ* hybridization reaction.
↪ *See* Chapter 4.

5.2.4.1 Tyramide

5.2.4.1.1 PRINCIPLE
See Figures 5.31 and 5.32.

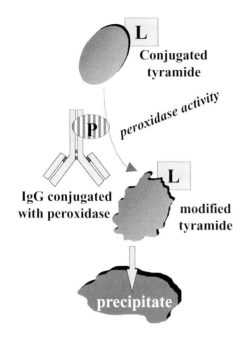

↪ Tyramide is a phenolic compound that precipitates after oxidation.
↪ Method of amplification of the immunological reaction that uses horseradish peroxidase as a label.
↪ Peroxidase (P) oxidizes tyramide conjugated to different labels (L).
↪ This leads to an accumulation of the label at the reaction site.
↪ The accumulation of the tyramide label is revealed by a classical reaction (*see* Section 5.2.2.2).

Figure 5.31 Diagram of the reaction.

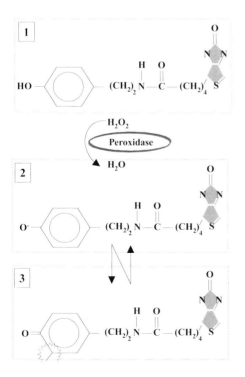

1 = Tyramide–biotin formula.

2 = Catalysis of tyramide–biotin by peroxidase.

3 = Activated tyramide–biotin.

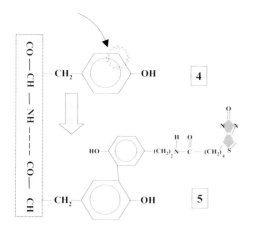

4 = Tyrosine from a protein.

5 = Formation of a diphenyl.

Figure 5.32 Formation of diphenyl compounds between tyramide–biotin and tyrosine.

5.2.4.1.2 TYRAMIDE CONJUGATES
See Figure 5.33

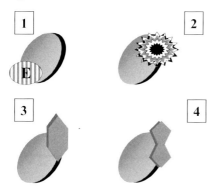

⇝ Tyramide can in most cases be conjugated to a label (e.g., an enzyme, fluorochrome, colloidal gold, or an antigenic molecule).
⇝ This principle can be applied to either a direct or an indirect reaction.

1 = Tyramide–enzyme: it is necessary to carry out a stained reaction.
2 = Tyramide–fluorochrome: direct visualization by fluorescence microscopy.
2 / 4 = Tyramide–antigen: for example, fluorescein, digoxigenin, biotin, dinitrophenyl, with revelation by an immunohistological reaction.
Figure 5.33 Tyramide conjugates.

5.2.4.1.3. ADVANTAGES/DISADVANTAGES
❏ *Advantages*
- High sensitivity of detection
- Very low level of background noise

- Rapid, simple reaction
- Multiple labeling

- Possibility of carrying out several amplification cycles
- Good level of resolution of detection

⇝ Useful in cases of low levels of expression
⇝ Accumulation of a large quantity of tyramide modified by peroxidase
⇝ < 30 min
⇝ The stained enzymatic reaction differs according to the substrate used (DAB, AEC, NBT – BCIP, Fast-Red).
⇝ With antigenic tyramide

⇝ Possibility of optimizing the resolution of detection by optimizing the concentrations of the antibody or the probe, or by varying the reaction time

❑ *Disadvantage*
- Endogenous peroxidase

↪ Inhibition of endogenous peroxidases (*see* Appendix B6.1.3)

5.2.4.2 Direct visualization of amplification

1. Summary of the different steps

↪ In the case of a fluorochrome conjugate

1 = Immunohistological reaction – Direct immunological detection (formation of the hapten–antibody complex).

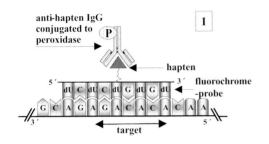

2 = Precipitation of tyramide–fluorochrome:

- Modification of conjugated tyramide by the action of peroxidase
- Precipitation

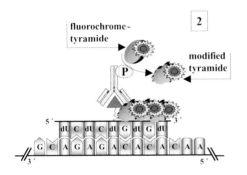

3 = Direct visualization of amplification – Observation by immunofluorescence

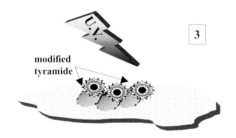

Figure 5.34 Amplification of a fluorochrome by catalytic deposition of tyramide–fluorochrome at reaction sites.

❑ *Advantages*

- Sensitivity

- Rapid method

↪ This is much more sensitive than the classical direct method (*see* Section 5.2.2.2).
↪ This is direct immunological reaction, then tyramide precipitation.

❏ *Disadvantage*
• Risk of background

⇝ It is necessary to reduce a specific reaction as far as possible (*see* Chapter 7)
⇝ The signal/background ratio depends on the presence of endogenous peroxidases and the concentration of the primary antibody, as well as the concentration of the modified tyramide and its incubation time.

2. Solutions
• Anti-hapten IgG conjugated to peroxidase

⇝ The peroxidase is necessary for the transformation of the tyramide.

• Buffers:
— Blocking buffer

⇝ 100 mM Tris–HCl buffer; 150 mM NaCl; 0.5% blocking agent; pH 7.6 (*see* Appendix B6.1.1)

— 1X amplification buffer
— Tris–HCl buffer
— 100 mM Tris–HCl buffer; 300 mM NaCl; 0.05% Tween 20; pH 7.6
• Tyramide–fluorescein

⇝ Commercial kit
⇝ *See* Appendix B3.6.1
⇝ Revelation buffer (*see* Appendix B3.6.5)

⇝ A commercial kit, or another fluorochrome (*see* Section 5.2.1.2.2)

3. Protocol — Direct visualization of the amplification revealed by a direct immunological reaction

a. Saturation of nonspecific sites:

⇝ All the following steps are carried out at room temperature.

• Blocking buffer	**15 to 30 min**
• Put the substrate in place	**≥ 100 μL/ section**

b. Incubation in the anti-digoxigenin antibody conjugated to peroxidase:

⇝ **1st step: formation of the hapten–antibody complex**
⇝ 100 mM Tris–HCl buffer; 300 mM NaCl; 0.05% Tween 20; pH 7.5

• Dilute in Tris–HCl / NaCl buffer	**1:50**
• Put the substrate in place	**≥ 80 μL/ section**
c. Incubate	**3 h at RT**

⇝ Moisture chamber (water on filter paper)
⇝ With shaking

d. Rinse with Tris–HCl/ NaCl/Tween 20 buffer	**3 × 5 min**

⇝ **2nd step: amplification**

e. Put the tyramine–fluorescein complex in place

• Dilute in a 1X amplification buffer	**1:50**
• Put the substrate in place	**≥ 20 μL/ section**
f. Incubate.	**10 min**

⇝ Moisture chamber (water on filter paper)

g. Rinse with Tris–HCl/ NaCl / Tween 20 buffer.	**RT** **3 × 5 min**	↪ With shaking
h. Mount in moviol.		↪ *See* Appendix B8.1.2
i. Visualize by fluorescence microscopy.		

5.2.4.3 Other possibilities for amplification

1. Detection of the hapten
   - Direct immunohistological reaction
   - Indirect immunohistological reaction
2. Precipitation of tyramide
3. Revelation of precipitated tyramide:
   - Tyramide conjugated to an enzyme
   - Tyramide conjugated to an antigen (e.g., biotin, digoxigenin, fluorescein)

↪ Possibility of amplification (*see* Section 5.2.3.2.3)
↪ Action of peroxidase on tyramide

↪ Stained reaction
↪ Amplification reaction
  • Streptavidin / biotin–label complex
  • Direct or indirect immunohistological reaction
↪ Increased risk of background

Amplification following tyramide precipitation remains difficult, and necessitates the development of a specific protocol for each application.

Chapter 6

Counterstaining and Modes of Observation

Contents

6.1	Counterstaining		191
	6.1.1	Classification of Stains	191
	6.1.2	Choice of Stain	192
	6.1.3	Counterstaining Protocols	193
		6.1.3.1 Toluidine Blue	193
		6.1.3.2 Cresyl Violet	193
		6.1.3.3 Harris's Hematoxylin	193
		6.1.3.4 Hematoxylin – Eosin	194
		6.1.3.5 Rapid Nuclear Red	194
		6.1.3.6 Methyl Green	194
6.2	Mounting Media		194
	6.2.1	Permanent	195
	6.2.2	Aqueous	196
	6.2.3	For Fluorescence	197
6.3	Modes of Observation		197
	6.3.1	Principle of the Light Microscope	197
		6.3.1.1 Equipment	197
		6.3.1.2 Formation of the Image and Resolution	198
		6.3.1.3 Enlargements	199
		6.3.1.4 Classification of Microscopes	199
	6.3.2	Bright Field	201
	6.3.3	Dark Field	202
	6.3.4	Epipolarization	203
	6.3.5	Fluorescence	204
6.4	Observation of the Signal		205
	6.4.1	Enzymatic Labels	205
	6.4.2	Choice of Mode of Observation	206

The aim of this step is to visualize tissue structures without modification of the *in situ* hybridization signal. The criteria for the choice of stain are summarized in Table 6.1.

↝ The color of the stain must be different from that of the signal.
↝ The stain must not be so intense as to mask the signal.

6.1 COUNTERSTAINING

6.1.1 Classification of Stains

Stains can be classified either from the chemical or the histological point of view. From the histological point of view, stains are divided into two groups:
- Natural stains
- Artificial stains

1. Natural stains

- Hematoxylin
- Carmine
- Orceine

↝ Natural stains are extracts of animal or vegetable origin.
↝ Hematoxylin is extracted from *Haematoxylum campechianum* (a Central American leguminous tree). The compound becomes a stain only after its oxidation and transformation into hemateine. The staining thus obtained is very stable.

2. Artificial (synthetic) *stains* are themselves divided into three groups:

a. The basic stains are nuclear stains, such as:

- Methylene blue

- Toluidine blue

- Methyl green

↝ They are classified according to their affinity for the nucleus and/or the cytoplasm.
↝ Their active component is a colored base. They are used in the form of salts whose base is colored, while its acid is colorless.
↝ Blue is the result of an oxido-reduction reaction.
↝ Very similar to methylene blue. It is metachromatic (in a neutral, alkaline, or acid solution). It stains the nucleus in a slightly acid solution.
↝ This is always used in an acid solution.
↝ It is important for fresh, nonfixed tissues.

		→ It stains the cytoplasm blue, and the nucleus a purplish red (used for nerve tissue).
	• Cresyl violet	
	b. Acid stains are cytoplasmic stains, e.g.:	→ Their active component is a colored acid. They are used in the form of a salt whose base is colorless, while its acid is colored.
	• Eosin	→ Derivatives of fluorescein. They are sodium or potassium salts of fluorescein bromide. The stain varies from very light pink to bright pink.
	• Light green	→ Color increases the contrast in black and white photographs.
	c. Neutral stains	→ The base and the acid are both colored.
	• Methylene eosinate blue solution	→ Or the May–Grünwald mixture (methylene eosinate blue dissolved in methyl alcohol) (used for smears)
	• Giemsa stain	→ There are different forms of Giemsa, e.g., a rapid form (used for smears) and a slow one (suitable for tissue sections).

6.1.2 Choice of Stain

Table 6.1 Detection of the Signal

Stains	Autoradiographic signal	Immunohistological signal					
		Alkaline phosphatase		Peroxidase			Colloidal gold
		NBT / BCIP	Fast - Red	DAB	4-Chloro-1-naphthol	AEC	Amplification
Toluidine blue	R						
Cresyl violet	R						R
Harris's hematoxylin			R	R		R	P
Eosin hematoxylin	P						P
Nuclear red			R		R		
Methyl green	R	R	R	R	R	R	R

R = recommended; P = possible

6.1.3 Counterstaining Protocols

6.1.3.1 Toluidine blue

↪ This is a cytoplasmic and nuclear stain.

1. On the preparation, place 0.02% toluidine blue in H$_2$O
 1 to 10 min
 RT

 ↪ See Appendix B7.1.
 ↪ The concentration varies from 0.02 to 0.25%.
 ↪ It can be diluted in oxalic acid neutralized by ammonia.

2. Rinse in distilled water.
 3 × 5 min

 ↪ It is possible to deepen the staining if necessary; in this case, go back to step 1.

3. Destain with alcohol 95%

 ↪ It is always possible to remove an excess of stain with alcohol 95%.
 ↪ Check with a microscope.

4. Mount according to the reaction.

 ↪ See Section 6.2.

5. Observe.

 ↪ See Section 6.3.

6.1.3.2 Cresyl violet

↪ Cresyl violet stains the nucleus purplish red and the cytoplasm blue.

1. On the preparation, place cresyl violet 1:10.
 1 to 5 min

 ↪ See Appendix B7.2; no overstaining.

2. Rinse with distilled water.
 30 s

 ↪ In the case of autoradiography, it is always possible to reduce the thickness of the gelatin layer in an acid water bath.

3. Mount according to the reaction.

 ↪ See Section 6.2.

4. Observe.

 ↪ See Section 6.3

6.1.3.3 Harris's hematoxylin

↪ This is a nuclear stain.

1. Over the preparation, pour Harris's hematoxylin
 30 s to 1 min

 ↪ See Appendix B7.4.

2. Wash in running water.
 5 min

3. Dehydrate with alcohol 95%, 100%
 2 min/bath

 ↪ It is possible to destain (alcohol/acid), but difficult after autoradiography.

4. Mount according to the reaction.

 ↪ See Section 6.2

5. Observe.

 ↪ See Section 6.3

6.1.3.4 Hematoxylin – eosin

1. Over the preparation, place:
   - Eosin   **30 s**   ⇰ *See* Appendix B7.3.
   - Running water   **1 min**   ⇰ Wash in water between stains.
   - Harris's hematoxylin   **30 s**   ⇰ *See* Appendix B7.4.
   - Running water   **1 min**

2. Mount according to the reaction.   ⇰ *See* Section 6.2.

3. Observe.   ⇰ *See* Section 6.3.

6.1.3.5 Rapid nuclear red

⇰ Not to be confused with Fast–Red.

1. Over the preparation, place rapid nuclear red.
                                                          1 min   ⇰ *See* Appendix B7.5.
2. Wash with running water.
                                                          1 min
3. Mount according to the reaction.   ⇰ *See* Section 6.2.
4. Observe.   ⇰ *See* Section 6.3.

6.1.3.6 Methyl green

1. Over the preparation, place methyl green.
                                           1 to 3 min   ⇰ *See* Appendix B7.6.
2. Wash with running water.
                                                          1 min
3. Mount according to the reaction.   ⇰ *See* Section 6.2.
4. Observe.   ⇰ *See* Section 6.3.

6.2 MOUNTING MEDIA

Mounting stabilizes the preparation in a stable medium between the slide and the coverslip, which should be of the same refractive index so as to give:
1. The best possible conditions for observation;
2. Conservation of the histological preparations and the reaction signal.   ⇰ Compatibility with the radioactive signal or the immunohistological reaction

There are two main methods:
- Permanent medium   ⇰ Dehydration and mounting in resin
- Aqueous medium   ⇰ Without dehydration

Table 6.2 Determination of the Choice of Mounting Medium

After Revelation	Permanent Mounting	Aqueous Mounting	Antifading Mounting
1. Immunohistological			
• NBT – BCIP	Possible	Recommended	
• Fast – Red		Recommended	
• DAB	Recommended	Possible	
• 4-Chloro-1-naphthol		Recommended	
• AEC		Recommended	
• Colloidal gold	Recommended	Possible	
2. Autoradiographic			
• On cryosections	Recommended	Possible	
• On paraffin sections	Recommended	Possible	
3. By immunofluorescence			
• On cryosections		Possible	Recommended
• On paraffin sections		Possible	Recommended

6.2.1 Permanent

The conventional mounting media (after dehydration) are permanent, nonaqueous mounts, available ready for use.

↪ *See* Appendix B8.2.

↪ For example, Acrytol, Clearium, Depex, Gurr, Eukitt, Mountex, Pertex Vitro-clud, (*see* Appendix B8.2)

↪ The classical mounting media are soluble in xylene and similar substances, and are used after immersion in an appropriate solvent (e.g., an aromatic solvent).

❑ *Advantages*
- Refractive index similar to that of glass
- Unlimited duration of conservation
- Stability of stains

❑ *Disadvantages*
- Requires dehydration

↪ Solubilization of some reaction precipitates (e.g., NBT–BCIP, 4-chloro-1-naphthol, AEC)

- Reversibility

↪ It remains sometimes possible to reverse a mounting by immersing the section in xylene (inversely proportional to the time that has elapsed since the mounting).

❑ *Protocol*
1. Dehydrate:
- Alcohol 70%, 95%, 100% 2×1 min/bath
- Xylene 2×2 min

↪ With shaking

↪ It is possible to prolong the duration of these two latter baths without risk.

2. Mount:
- Place a drop of the medium on a slide that has been given a light coating of xylene (for better spreading).
- Put the coverslip in place immediately.
- Eliminate bubbles.

⇝ Mounting medium (Permount, Eukitt, Depex, etc.)

⇝ Use a heating block for better spreading.

3. Observe.

⇝ *See* Section 6.3.

6.2.2 Aqueous

⇝ *See* Appendix B8.1

Aqueous mounts are necessary in the case of stained or nonstained sections if the reaction precipitate is soluble in alcohol. These are:

- Commercial aqueous mounting media

⇝ For example, Aquamount, Aquateck, Crystal mount, Glycergel (*see* Appendix B8.1).

- Buffered glycerine

⇝ *See* Appendix B8.1.1.

- Moviol

⇝ *See* Appendix B8.1.2.

❏ *Advantages*
- Conservation of hydrosoluble precipitates

- Reversibility

⇝ A second reaction can be carried out after the first has been photographed.

- Conservation of fluorescence

⇝ Add an "antifading" agent.

❏ *Disadvantages*
- Refractive index similar to that of glass

- Conservation limited

⇝ Preferably at 4°C

- Choice of counterstaining

⇝ Stable in a hydrophilic medium

❏ *Protocol*
1. Stop the reaction with distilled water.

2. Mount:
- Deposit a drop of the medium.
- Put the coverslip in place.
- Eliminate bubbles.

3. Observe.

⇝ *See* Section 6.3.

6.2.3 For Fluorescence

With an antifading additive, some mounting media can be used for conservation, and others for the amplification of fluorescence.

Some aqueous mounts can also be used, given that they stabilize and amplify fluorescence.

❑ *Protocol*
1. Rinse with buffer.
   3 × 5 min
2. Mount:
   - Deposit a drop of the medium
   - Put the coverslip in place
   - Eliminate bubbles

3. Observe by fluorescence microscopy.

↪ *See* Appendix B8.3.

↪ For example, Mountex, Pertex, Vitro-clud.

↪ For example, 90% glycerol, or Moviol (*see* Appendix B8.1).

↪ *See* Section 6.3.5.

6.3 MODES OF OBSERVATION

6.3.1 Principle of the Light Microscope

6.3.1.1 Equipment

A light microscope is made up of the following components:
1. A source of radiation and a means of illuminating the object under examination;
2. A condensing lens, situated under the object, which focuses a cone of light rays and directs it toward the slide;
3. A slide, situated between the condensing and objective lenses, which can move in two directions;
4. An objective lens, which receives a beam of light rays emitted by the sample, and enlarges the image;

5. An eyepiece, which enlarges direct vision;

↪ *See* Figure 6.1.

↪ The type of radiation determines the interaction with the object.

↪ Numerous parameters can be introduced into the condenser (dark field, polarizer, phase contrast, etc.).

↪ Motorized plates make possible computer-controlled movement.

↪ This is the most important component.
↪ The objective-holder has different objective lenses (magnification 4 to 100×)
↪ Its quality (digital aperture) will determine the resolution of the photographic image.

↪ The eyepiece is attached to the upper part of the microscope, and is generally binocular.
↪ It is not involved in photographic enlargements.

6. Two focusing knobs, for coarse and fine (micrometric) focusing.

⇝ The knobs are located on the frame.

The different parts are attached to a relatively heavy stand.

⇝ The stand gives stability for photography.

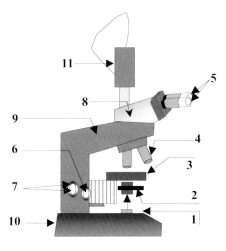

1 = Light source
2 = Condensing lens
3 = Slide
4 = Objective lens
5 = Eyepieces
6 = Condensing lens adjustment knobs
7 = Coarse and fine adjustment knobs
8 = Body
9 = Frame
10 = Stand
11 = Photographic system

Figure 6.1 Diagram of a light microscope.

6.3.1.2 Formation of the image and resolution

⇝ *See* Figure 6.2. Image is virtual, magnified, and inverted.

The objective must produce a clear, enlarged image. The power of a lens depends on its focal length (*see* Figure 6.2).

⇝ The focal length is the distance between the center of the lens and the focal point (this being the point of convergence of rays parallel to the axis of the lens).

The enlargement produced by the objective is expressed by the ratio A'B' / AB:

⇝ *See* Figure 6.2. It depends essentially on the quality of the optics.

• A powerful objective lens (L2) has a shorter focal length than a weak one (L1): its resolving power is larger, and its working distance shorter.

⇝ The distance between the center of the lens (L) and the image (AB) is called the focal length (distance F2 < F1).

⇝ Resolution is the minimal distance required between two objects so that they will appear distinct. It is determined by the formula $R = 0.61\ \lambda/NA$, where λ is the wavelength and NA the numerical aperture of the objective lens.

• The angle of the luminous cone (α) is larger with a powerful objective lens than with a weak one.

⇝ L2 has a larger angular opening ($\alpha 1 > \alpha 2$).

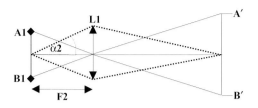
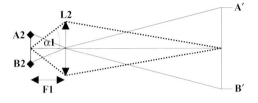

AB = sample

A'B' = image of the sample

L = objective lens

F = focal length

α = angle of the luminous cone

Figure 6.2 Formation of the image in light microscopy.

6.3.1.3 Enlargements

The enlargement of a microscopic observation is calculated by multiplying the enlargement of the objective by that of the eyepiece.

↪ Example of the enlargement of an observation: if a 45× objective lens is used with a 10× eyepiece, the final enlargement will be 450×.

Photographic enlargement corresponds to the enlargement of the objective lens multiplied by that of the photographic system and that of the projection system.

↪ The projection system can be that of a printout on paper, that of a projector and a screen, or that of a computing system.

The final observed enlargement is a net result of numerous factors, which it is necessary to indicate.

↪ The enlargement can be given by a scale on the photograph.

6.3.1.4 Classification of microscopes

The classification of microscopes is based on the nature of the radiation. Light microscopy uses either electromagnetic radiation in the visible spectrum or ultraviolet radiation close to the wavelength of visible light.

↪ The wavelength of visible light is between 0.4 μm (violet) and 0.7 μm (red).
↪ See Figure 6.3.

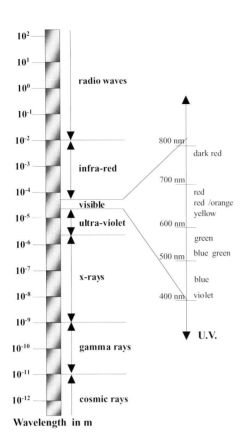

Figure 6.3 Radiation scale.

Several modes of light microscopy have been developed:

⇒ See Table 6.4.

- Bright field

⇒ For enzymatic reactions, with colloidal gold and autoradiography

- Dark field

⇒ Used with autoradiography and colloidal gold

- Epipolarization

⇒ For colloidal gold reactions latensified by silver salts and autoradiography

- Fluorescence

⇒ For reactions using fluorochromes

- Phase contrast / interferential

⇒ Living cells, or visualization of cells without staining

6.3.2 Bright Field

1. Principle — The light passes through the object before reaching the eye. The background of the preparation is light colored, with the object absorbing some of the transmitted light. The image formed is more or less dark (according to the stain used) on a bright background (Figure 6.4).

↪ The conventional light microscope is known as the "bright-field" microscope.
↪ The color(s) of the preparation depend on the wavelengths absorbed by the stain(s) present.

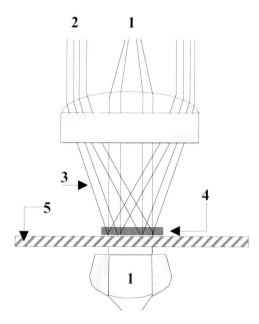

1 = Direct light

2 = Light beam

3 = Diffracted light
4 = Objective lens
5 = Sample

Figure 6.4 Diagrammatic representation of the beams of light that pass through the preparation (illumination by transmitted light).

2. Utilization — The bright-field mode is a standard for all forms of microscopy. It is used in the observation of tissue structures.

↪ In autoradiography, the silver grains appear black.
↪ This allows visualization of the chromogen precipitate.
↪ In fluorescence, it allows structures to be matched to the visualization of the signal (associated with phase contrast).

❏ *Advantages*
- Simple to use
- Visualization of signals obtained by immunohistochemistry and autoradiography

❏ *Disadvantages*
- Little contrast

- Requires the observed structures to be stained
- Low level of definition

↪ Only the contours of the object appear (e.g., the colloidal gold label gives a reddish staining).

↪ The particles are small, and thus difficult to discern.

Counterstaining

6.3.3 Dark Field

1. Principle— See Figure 6.5.

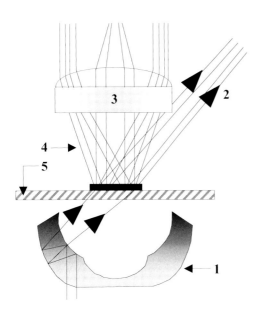

2. Utilization — The dark-field microscope is generally used in autoradiography to observe metal particles (silver grains or colloidal gold) in nonstained preparations.

❏ *Advantages*
- Usable for low-magnification enlargements

- Discrimination between light-reflecting particles and tissue

- Characterization of silver grains by autoradiography

- Possibility of quantitative methods for counting grains

❏ *Disadvantages*
- Requires a specific system

- Applications limited to low levels of magnification

- Requires double photography (dark field and phase contrast)

↪ Illumination by transmission with a dark field

↪ The difference between the dark-field microscope and the bright-field microscope essentially concerns the relationship between the illumination aperture and the observation aperture.

(1) Dark-field condensing lens: This hollow set of mirrors creates an illumination of large aperture, in such a way that the light transmitted directly through the specimen **(2)** is not collected by the **objective lens (3)**. Only **the light diffracted (4)** by the sample **(5)** reaches the objective lens. The dark field is produced by a beam of incident light that is weaker than with a bright field.

Figure 6.5 Diagrammatic representation of the beams of light that pass through the preparation (illumination with dark field).

↪ The silver grains appear bright against a dark field.

↪ Useful in cases of weak labeling.

↪ Distinction between metallic silver and tissue pigment

↪ Integration of reflected light

↪ Dark-field condensing lens (and generally a phase contrast)

↪ To localize the signal on the tissue

6.3.4 Epipolarization (Epi-illumination)

1. **Principle** — The use of polarized light means that parasitic light due to reflections from the optical components can be excluded (Figure 6.6).

⇝ Combines diffraction and transmission with a bright field

⇝ Natural light vibrates in a random way in all directions in the course of its propagation. It is said to be polarized when it vibrates in just one particular direction.

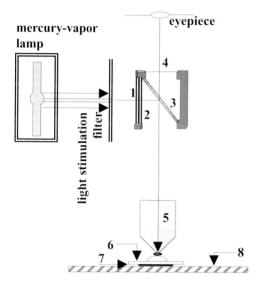

1 = UV filter
2 = Polarizer
3 = Dichroic mirror
4 = Analyzer

5 = Objective lens (a condensing role and a collecting role) — immersion oil
6 = Object glass
7 = Object
8 = Slide

Figure 6.6 Epi-illumination.

2. **Utilization** — Observation in epipolarization is especially used:
• After autoradiographic revelation, with the silver grains appearing bright against a stained section background;
• After immunological revelation with colloidal gold latensified by silver salts.

⇝ Radioactive probes

⇝ Antigenic probes

❏ *Advantages*
• Increased sensitivity / increased contrast
• Simultaneous visualization of the signal and the tissue
• Attenuation of the background noise

• Excellent quality of images

❏ *Disadvantages*
• Requires special equipment
• Requires very precise adjustment of the microscope
• Difficult to use with low levels of enlargement

⇝ Useful in cases of weak labeling
⇝ Combination of dark field and transmitted light
⇝ Possibility for weak-intensity signals to be attenuated by the transmitted light
⇝ Black and white, and especially color

⇝ Epi-illumination, filters, transmitted light
⇝ Useful in cases of weak labeling

⇝ Dark field preferable

6.3.5 Fluorescence

1. *Principle* — The fluorescence microscope is a conventional microscope in which the light source is a mercury-vapor lamp combined with narrowband filters. The emission spectrum is very wide.

↬ The most common type uses epifluorescence. Unlike the bright-field technique, where the light passes through the object, UV radiation illuminates the preparation through the objective lens, without passing through the glass slide, and the radiation emitted by the excited fluorochrome is observed through the objective lens.

↬ The most suitable type of radiation is provided by mercury-vapor lamps, in that it possesses the required energy.

There are two principles:
- Transmitted fluorescence

↬ UV radiation passes through the preparation, as in the case of the bright-field microscope.

- Epi-fluorescence

↬ The UV radiation passes through the objective lens before reaching the preparation, but does not pass through the slide, which would absorb a large proportion of it.
↬ The eye receives no UV radiation.
↬ According to the wavelength of excitation of the fluorochrome used (*see* Section 5.2.1.2.2 and Table 5.1).

The sample is illuminated by UV light that has passed through an excitation filter, which lets through only light of the desired wavelength (Figure 6.7).

↬ In normal conditions, molecules are in a certain vibratory state, associated with a determinate level of energy. If the molecule is struck by energetic radiation, i.e., in this case, UV, its vibratory state is modified, as well as its energy. The aromatic nuclei of the molecule, and its electrons, vibrate more strongly, they are "excited." The molecule returns to its equilibrium state by emitting a photon with a shorter wavelength than that of the exciting photon.

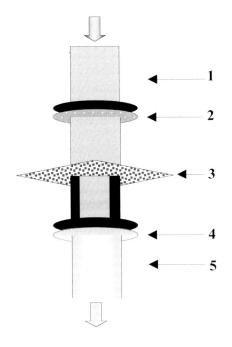

1 = UV
2 = UV selector filter
3 = Preparation
4 = Separator filter
5 = Fluorescence

Figure 6.7 Excitation by UV.

There exists a range of excitation filters for the UV spectrum.

Complementary filters positioned behind the objective lens and eyepiece make it possible to exclude undesirable fluorescence signals.

2. *Utilization* — The preparation must be protected from all light of exciting wavelength, and must be observed very rapidly.

❏ *Advantages*
- High sensitivity

- Images with excellent contrast

- Multiple detections (selectivity)

- Quantification

❏ *Disadvantages*
- Necessary to observe and analyze rapidly

- Requires special equipment

↝ Each filter corresponds to a particular fluorochrome (*see* Section 5.2.1.2.2 and Table 5.1).

↝ Intensity

↝ Light on a dark field, or a fluorescent counterstaining at another wavelength

↝ For a given preparation, possible to observe the fluorescence of several labels by changing filters, or by using multiband filters

↝ Integration of the light emitted by fluorochromes

↝ Very short duration of the fluorescent emission.

6.4 OBSERVATION OF THE SIGNAL

6.4.1 Enzymatic Labels

Table 6.3 Recapitulation of the Different Enzymes and Their Chromogens

Enzymes	Chromogenic Substrates	Precipitates
Alkaline phosphatase	NBT – BCIP	Blue
	Fast – Red	Red
Peroxidase	DAB	Brown
	AEC	Red
	4-Chloro-1-naphthol	Blue-violet

Counterstaining

6.4.2 Choice of Mode of Observation

Table 6.4

Labels	Revelation	Modes of Observation			
		Bright Field	Dark Field	Epipolarization	Fluorescence
Enzymes	Chromogens	Recommended, ±s			
Fluorochromes	Without				Recommended
Colloidal gold	Latensified with silver	Possible	Recommended, − s	Recommended, + s	
Radioisotopes	Latensified with silver	Recommended, + s	Recommended, − s	Recommended, + s	

+ s, with staining; − s, without staining; ± s, with or without staining.

Chapter 7

Controls and Problems

Contents

7.1	Signal/Background Ratio	211
	7.1.1 Probes	211
	7.1.2 Samples	212
	7.1.3 Pretreatments	212
	7.1.4 Hybridization	212
	7.1.5 Washing	213
	7.1.6 Revelation	213
	7.1.7 Observation	213
7.2	Sensitivity/Specificity	214
	7.2.1 Probes	214
	7.2.2 Samples	214
	7.2.3 Pretreatments	215
	7.2.4 Hybridization	215
	7.2.5 Washing	216
	7.2.6 Revelation	216
	7.2.7 Observation	216
7.3	Controls	217
	7.3.1 Probes	217
	7.3.2 Labeling	217
	7.3.3 Tissue	217
	7.3.4 Hybridization, Washing	218
	7.3.5 Revelation	218
	7.3.6 Variation of the Signal	218
	7.3.7 Verification by Other Techniques	218
7.4	Recapitulative Table	219

7.1 SIGNAL / BACKGROUND RATIO

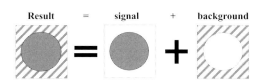

↝ The observed result is a combination of the signal and the background. Only the relationship between these two parameters enters into the final analysis

7.1.1 Probes

Criteria

1. Nature
   - cDNA → → ↝ Reference
   - Single-stranded DNA ↗ →
   - Oligonucleotides **Several** ↝ Multiplication of the number of oligonucleotides for a given target
   - cRNA ↗ ↘ ↝ Due to a specific high level of activity, and the stability of RNA–RNA hybrids
2. Homology
   - 100% → → ↝ Reference
   - <100% ↘ ↗ ↝ Loss of stability of hybrids and nonspecific bonds
3. Label
   - Radioactive ↗ ↗
   - Antigenic ↘ ↘ ↝ Threshold effect for detection
4. Specific activity
   - + ↘ ↘ ↝ Threshold effect
   - ++ → → ↝ Reference
   - +++ ↗ ↗ ↝ Possibility of stringent washings
5. Purification ↗ ↘ ↝ Decrease in nonspecific bonds

7.1.2 Samples

Criteria

1. Sampling conditions — ↗ — ⇒ Decrease in exogenous RNases
2. Preservation of nucleic acids — ↗ — ⇒ Specific hybridizations

7.1.3 Pretreatments

Criteria

1. Fixation
   - \+   ↗   ↘   ⇒ Preservation of the accessibility of nucleic acids
   - +++ ↘   ↗   ⇒ Formation of nucleic acid–protein complexes
2. Deproteinization   ↗   ↘   ⇒ Facilitates accessibility to nucleic acids
3. Acetylation   ↘   ↘   ⇒ Blockage of NH$_2$ functions
4. Prehybridization   ↘   ⇒ Saturation of nonspecific sites
5. Denaturation
   - 0   0   ↗   ⇒ If the target is a DNA, the signal must be absent.
   - \+   →   ⇒ Accessibility of double-stranded targets
6. Storage
   - Without H$_2$O   ⇒ Inhibition of RNase activity
   - Without precautions ↘ ⇒ Activation of RNases

7.1.4 Hybridization

Criteria

1. Probe concentration
   - \+   ↘   ↘   ⇒ Absence of saturation of targets
   - ++   ↗   ⇒ Saturation of targets
   - +++   ↗   ⇒ Increase in the possibilities of nonspecific bonds
2. NaC concentration ≈ 300 mM   ↗   ↗   ⇒ Facilitation of hybridization
3. Dextran sulfate concentration   ↗   ⇒ Probe concentration
4. tRNA, DNA concentrations   ↘   ⇒ Saturation of nonspecific bonds
5. Detergents   ↘   ⇒ Reduction of hydrophobic regions
6. DTT   ↘   ⇒ Reduction of the oxidation of ^{35}S

7.1 *Signal / Background Ratio*

7. Formamide concentration	→	↘	↪ Increase in the specificity of hybridization
8. Temperature			
• +	↘	↗	↪ Possibility of nonspecific hybridizations
• ++	↗	→	↪ Reduction in nonspecific hybridizations
• +++	↘	↘	↪ Possibility of denaturation
9. Time			
• <3 h	↘	↘	↪ Incomplete hybridization
• 3 h	↗		↪ Sufficient hybridization
• >3 h	→	↗	↪ Risk of an increase in nonspecific hybrids

7.1.5 Washing

Criteria

1. NaCl concentration			
• +++		↗	↪ Stability of nonspecific hybrids
• +	↗	↘	↪ Denaturation of nonspecific hybrids
2. Temperature			
• +	↗	↗	↪ Stability of hybrids
• +++	↘	↘	↪ Denaturation of hybrids
3. Time	↘	↘	↪ Labilization of hybrids

7.1.6 Revelation

Criteria

1. Autoradiographic	↗	↗	↪ Sensitivity
2. Immunohistological	↘	↘	↪ Detection threshold
• Direct method	↘	↗	↪ Nonspecific adsorption
• Indirect method	↗	→	↪ Increase in sensitivity
• Amplification	↗↗	↗	↪ Lowering of the detection threshold

7.1.7 Observation

Criteria

1. Bright field	→	→	
2. Dark field	↗	↗	↪ Amplification of the signal
3. Fluorescence	↘	↘	↪ Presence of a threshold
4. Epipolarization	↗	↘	↪ Attenuation of a diffuse signal
5. Confocal microscope	↗		↪ Lowering of the detection threshold
6. Image analysis	↗↗	↘↘	↪ Signal analysis, quantification

7.2 SENSITIVITY / SPECIFICITY

7.2.1 Probes

Criteria

	sensitivity	specificity	
1. Type			
• cDNA	↘	→	⇨ Rehybridization of complementary strands
• Single-stranded DNA	→	→	⇨ Hybridization of probe with its target
• Oligonucleotide	→	→	⇨ Hybridization of probe with its target
• Oligonucleotides	↗	→	⇨ Increase in potential hybrids
• cRNA	↗	→	⇨ High specific activity, stability of hybrids
2. Homology			
• 100%	→	→	⇨ Optimal conditions
• <100%	↘	↘	⇨ Risk of nonspecific hybrids
3. Label			
• Radioactive	↗	→	⇨ No detection threshold
• Antigenic	↘	↘	⇨ Detection threshold, endogenous antigen
4. Specific activity			⇨ To check probe labeling
• +	↘	→	⇨ Detection threshold
• ++	→	→	⇨ Saturating concentration
• +++	↗	↘	⇨ Increase in nonspecific signal
5. Purification	↗	↗	⇨ Decrease in nonspecific reactions

7.2.2 Samples

Criteria

	sensitivity	specificity	
1. Condition for sampling	↗	↗	⇨ Limitation of contamination
2. Preservation of nucleic acids	↗	↗	⇨ Limitation of destruction

7.2.3 Pretreatments

Criteria

1. Fixation
   - + ↗ ↘ ⇒ Preservation, but potential diffusion, of targets
   - +++ ↘ ⇒ Blockage of nucleic acid–protein complexes
2. Deproteinization ↗ ⇒ Accessibility of target nucleic acids
3. Acetylation ↘ ⇒ Blockage of NH_2 functions
4. Prehybridization ↗ ⇒ Blockage of nonspecific sites
5. Denaturation
   - − 0 ⇒ Indispensable with DNA targets
   - + → → ⇒ Accessibility of the target
6. Storage
   - Without H_2O → → ⇒ Conservation of targets
   - Without precaution ↘↘ ⇒ Loss of targets

7.2.4 Hybridization

Criteria

1. Probe concentration
   - + ↘ ↗ ⇒ Not all targets hybridized
   - ++ ↗ ⇒ All targets hybridized
   - +++ ↘ ⇒ Increase in nonspecific bonds
2. NaCl concentration ↗ ⇒ Stabilization of nucleic acid–nucleic acid hybrids
   ≈ 300 mM
3. Dextran sulfate concentration ↗ ⇒ Probe concentration
4. tRNA, DNA concentrations ↗ ⇒ Saturation of nonspecific sites
5. Detergents ↗ ⇒ Denaturation of nonspecific sites
6. DTT ↗ ↗ ⇒ Protection of the label
7. Formamide concentration ↗ ↗ ⇒ Facilitation of hybridization
8. Temperature
   - + ↗ ↘ ⇒ Facilitation of hybridizations
   - ++ → → ⇒ Optimal hybridization
   - +++ ↘ ↗ ⇒ Denaturation of hybrids
9. Time
   - <3 h ↘ ↗ ⇒ Hybridization of some specific hybrids
   - 3 h → ↗ ⇒ Hybridization of all the specific hybrids
   - >3 h ↗ ↘ ⇒ Hybridization beyond the specific hybrids

7.2.5 Washing

Criteria

	sensitivity	specificity	
1. NaCl concentration			
• +++	→	↗	⇝ Stability of nucleic acid–nucleic acid hybrids
• +	↘	↘	⇝ Instability of hybrids
2. Temperature			
• +	↗	↘	⇝ Weak denaturation of nonspecific hybrids
• +++	↘	↗	⇝ Denaturation of nonspecific more than specific hybrids
3. Time	↘	↘	⇝ Hydrolysis of hybrids

7.2.6 Revelation

Criteria

	sensitivity	specificity	
1. Autoradiographic	↗	↘	⇝ Absence of a detection threshold, autoradiographic background
2. Immunohistological	↘	↗	⇝ Existence of a detection threshold
3. Methods			
• Direct	↘	↘	⇝ High detection threshold, background
• Indirect	↗	↗	⇝ Lower threshold
• Amplification	↗	↘	⇝ Risk of background

7.2.7 Observation

Criteria

	sensitivity	specificity	
1. Bright field	→	→	⇝ Reference
2. Dark field	↗	↘	⇝ Excellent visualization of the autoradiographic signal
3. Fluorescence	↘	→	⇝ Detection threshold
4. Epipolarization	↗	↗	⇝ Attenuation of background
5. Confocal microscope	↗		⇝ Spatial visualization of the signal
6. Image analysis	↗↗	→	⇝ Signal analysis, quantification

7.3 CONTROLS

7.3.1 Probes

Criteria

1. Search for homologous sequences in a gene bank
   ↝ Risk of the existence of sequences with strong homology, which may cause the formation of hybrids
2. Hybridization with a "sense" probe ✓
   ↝ For oligonucleotides and RNA probes
3. Hybridization with a heterologous probe ✓
   ↝ For all types of probes
4. Probe competition ✓
   ↝ Between the labeled probe and a large quantity of the same, but nonlabeled, probe: negative hybridization
5. Nondenaturation of a cDNA probe ✓
   ↝ Verification of the reactivity of the tissue being studied

7.3.2 Labeling

Criteria

1. Radioactive
   ↝ Determination of specific activity
2. Antigenic
   ↝ Dot blot on membrane

7.3.3 Tissue

Criteria

1. Positive tissue ✓
   ↝ Tissue or cells known to express the target nucleic acid: positive hybridization
2. Negative tissue ✓
   ↝ Tissue or cells known not to express the target nucleic acid: negative hybridization
3. Destruction of the target by enzymatic treatment:
   • DNase ✓
   ↝ Degradation: hybridization < 0
   • RNase ✓
   ↝ Fragmentation of RNA targets (hybridization < 0), which must be followed by numerous washings

7.3.4 Hybridization, Washing

Criteria

1. Hybridization with a heterologous probe of the same type ✓ → Verification of the reactivity of the tissue being studied
2. Hybridization in the absence of a probe ✓
3. Action on stringency → Variation of the NaCl concentration, or of the hybridization temperature and/or washing: modification of the signal

7.3.5 Revelation

Criteria

1. Autoradiographic ✓ → Determination of background
2. Immunohistological
   • Inhibition ✓ → Search for endogenous activity:
      • Endogenous biotin
      • Endogenous peroxidase
      • Endogenous alkaline phosphatase

7.3.6 Variation of the Signal

Criteria

1. Experimental variations in the expression of the target nucleic acid → Determination of background
2. Quantification of the signal

7.3.7 Verification by Other Techniques

Criteria

1. Dot blot, Southern blot, Northern blot → Demonstration of the presence of the target nucleic acid in tissue
2. PCR → Determination of the presence or absence of the target nucleic acid
3. *In situ* PCR → Amplification of the signal
4. Immunocytology → Revelation of the corresponding protein, starting with the target nucleic acid

7.4 RECAPITULATIVE TABLE

Criteria		Signal	Background	Sensitivity	Specificity
Probes					
1. Type	• cDNA	→	→	↘	→
	• Single-stranded DNA	↗	→	↗	→
	• Oligonucleo-tides	x oligo		→	→
	• cRNA	↗	↘	↗	→
2. Homology	• 100%	→	→	→	→
	• <100%	↘	↗	↘	↘
3. Label	• Radioactive	↗	↗	↗	→
	• Antigenic	↘	↘	↘	↘
4. Specific activity	• +	↘	↘	↘	→
	• ++	→	→	→	→
	• +++	↗	↗	↗	↘
5. Purification		↗	↘	↗	↗
Sampling					
1. Sampling conditions		↗		↗	↗
2. Preservation of nucleic acids		↗		↗	↗
Pretreatments					
1. Fixation	• +	↗	↘	↗	↘
	• +++	↘	↗	↘	
2. Deproteinization		↗	↘	↗	
3. Acetylation		↘	↘	↘	
4. Prehybridization			↘		↗
5. Denaturation	• 0	0	↗	0	
	• +	→		→	→
6. Storage	• Without H₂O			→	→
	• Without precaution	↘		↘↘	

		Signal	Background	Sensitivity	Specificity
Hybridization					
1. Probe concentration	• +	↘	↘	↘	↗
	• ++	↗		↗	
	• +++		↗		↘
2. NaCl concentration	• ≈ 300 mM	↗	↗	↗	
3. Dextran sulfate concentration		↗		↗	
4. Concentrations of tRNA, DNA, etc.			↘		↗
5. Detergents, etc.			↘		↗
6. DTT			↘	↗	↗
7. Formamide concentration		↗	↘	↗	↗
8. Temperature	• +		↗	↗	↘
	• ++	↗		→	→
	• +++	↘		↘	↗
9. Time	• <3 h	↘	↘	↘	↗
	• 3 h	↗		→	↗
	• >3 h	→	↗	↗	↘
Washing					
1. NaCl concentration	• +++		↗	→	↗
	• +	↗	↘	↘	↘
2. Temperature	• +	↗	↗	↗	↘
	• +++	↘	↘	↘	↗
3. Time		↘	↘	↘	↘
Revelation					
1. Autoradiography		↗	↗	↗	↘
2. Immunohistology		↘	↘	↘	↗
	• Direct method	↘	↗	↘	↘
	• Indirect method	↗	→	↗	↗
	• Amplification	↗↗	↗	↗	↘
Observation					
1. Bright field		→	→	→	→
2. Dark field		↗	↗	↗	↘
3. Fluorescence		↘	↘	↘	→
4. Epipolarization		↗	↘	↗	↗
5. Confocal microscopy		↗		↗	
6. Image analysis		↗↗	↘↘	↗↗	→

Chapter 8

Typical Protocols

Contents

8.1	Preparation of Probes		225
	8.1.1	Radioactive cDNA Probe	225
	8.1.2	Radioactive Oligonucleotide Probe	226
	8.1.3	Radioactive cRNA Probe	227
	8.1.4	Antigenic cDNA Probe	228
	8.1.5	Antigenic Oligonucleotide Probe	230
	8.1.6	Antigenic cRNA Probe	230
8.2	Preparation of Tissue		232
	8.2.1	Frozen Tissue	232
	8.2.2	Paraffin-Embedded Tissue	233
	8.2.3	Preparation of Cells	234
		8.2.3.1 In Suspension	234
		8.2.3.2 In Culture	234
		8.2.3.3 Fixation	234
8.3	Pretreatments		235
8.4	Radioactive Probes: Hybridization, Washing		236
	8.4.1	Radioactive cDNA Probe	236
	8.4.2	Radioactive Oligonucleotide Probe	238
	8.4.3	Radioactive cRNA Probe	239
8.5	Autoradiography		241
	8.5.1	Macroautoradiography	241
	8.5.2	Microautoradiography	241
	8.5.3	Autoradiographic Revelation	241
	8.5.4	Counterstaining	241
	8.5.5	Mounting after Dehydration	242
	8.5.6	Modes of Observation	242
8.6	Antigenic Probes: Hybridization, Washing		243
	8.6.1	Antigenic cDNA Probe	243
	8.6.2	Antigenic Oligonucleotide Probe	244
	8.6.3	Antigenic cRNA probe	245
8.7	Immunocytological Detection		247
	8.7.1	Immunohistological Reaction	247
	8.7.2	Revelation with Alkaline Phosphatase	247
	8.7.3	Revelation with Peroxidase	248
	8.7.4	Counterstaining	249
	8.7.5	Mounting	249

8.1 PREPARATION OF PROBES

↪ *See* Chapter 1.

8.1.1 Radioactive cDNA Probe

↪ **Gloves must be worn.**
↪ Labeling by nick translation

Reactive medium
Precipitation

1. Reactive medium
a. To the dry pellet, add:
- Nick buffer 10X   5 µL
- Nonlabeled dNTP   3 µL / nucleotide
  2 nM/µL
- Plasmid DNA   200 ng
- Enzymes   5 µL
- α^{35}S-dCTP   50 µCi
- Sterile water   to 50 µL

↪ If ^{35}S–dCTP, use the dGTP, dTTP, dATP mix
↪ Concentration 100 to 500 ng/µL

↪ ≈ 1000 µCi/mmol (α^{35}S–dNTP)

b. Vortex, centrifuge.   ≈1000 rpm at 4°C
↪ To recover all the liquid at the bottom of the tube

c. Incubate.   75 to 90 min at 15°C
↪ Incorporation of nucleotides (all the missing bases are replaced)

d. Add phenol/chloroform.   50 µL
↪ **Optional**
↪ To stop the reaction; also possible to stop the reaction by heating for 10 min at 65°C

e. Recover the upper aqueous phase.
↪ Draw off 1 µL to measure total radioactivity before the extraction of the probe.

2. Precipitation
a. Add:
- tRNA 10 mg/mL   6 µL
- DNA 10 mg/mL   7 µL

b. Precipitate with alcohol
- 3 M sodium acetate   7 µL
- Ethanol 100% at –20°C   140 µL

↪ To eliminate free nucleotides

c. Incubate.   30 to 60 min at –80°C, or overnight at –20°C
↪ Precipitation

d. Centrifuge.   ≥ 14,000 g 30 min at 4°C

e. Wash.
- Ethanol 70% at –20°C   100 µL 3 × 1 min
↪ Without resuspending the pellet

f. Remove the supernatant.
↪ Orient the tube

g. Dry the pellet.

h. Dilute with water.
↪ Or TE buffer (*see* Appendix B3.51)

8.1.2 Radioactive Oligonucleotide Probe

⇝ Gloves must be worn.
⇝ 3' extension labeling

Reactive medium

Precipitation

1. Reactive medium

a. Place the following reagents in a sterile Eppendorf tube, in the indicated order:

• 10 to 100 pmol oligonucleotides	2 µL
• Buffer 5X	2 µL
• 25 mM CoCl$_2$	1 µL
• Labeled nucleotide	50 µCi
• TdT 25 units/µL	1 µL
• Sterile distilled water	to 10 µL

⇝ Potassium cacodylate 1 M; 125 mM Tris–HCl; BSA 1.25 mg / mL

⇝ ≈1000 µCi / mmol (α^{35}S-dNTP)
⇝ Terminal deoxytransferase

b. Incubate. **60 min at 37°C**

2. Precipitation

a. Add 10 mg/mL tRNA. **2 µL**

⇝ Nucleic acid of large size (carrier), which facilitates precipitation

b. Precipitate in alcohol:
• 7.5 M ammonium acetate	5 µL
• Ethanol 100% at –20°C	36 µL

⇝ To eliminate free nucleotides
⇝ 1:10 of final volume

⇝ Two to three times the reactive volume to be precipitated

c. Incubate. **30 to 60 min at –80°C or overnight at –20°C**

⇝ Precipitation

d. Centrifuge. **≥14,000 g 30 min at 4°C**

e. Remove the supernatant.

⇝ Orient the tube

f. Dry the pellet.

8.1.3 Radioactive cRNA Probe

In vitro transcription

Digestion of the DNA

Purification

Precipitation

⇝ Gloves must be worn.

⇝ Labeling by *in vitro* transcription

1. In vitro *transcription*

a. Place the following reagents in an Eppendorf tube, in ice, in the indicated order:

⇝ Riboprobe labeled by α^{35}S-UTP

• Linearized plasmid	**1 µg**
• Transcription buffer 5X	**2 µL**
• 100 mM DTT	**2 µL**
• RNasin 20 U/µL	**1 µL**

⇝ ^{35}S protection
⇝ Final concentration 1 U/µL
⇝ RNase inhibitor, to avoid the degradation of the RNA synthesized; this reagent is not resistant to high temperatures; must be added at the last moment

• rNTP mix: — 10 mM ATP — 10 mM GTP — 10 mM CTP	**2 µL**

⇝ Final concentration 0.4 mM
⇝ First, prepare a mixture of all the components, but without DNA and RNA polymerase, and neutralize at pH 7.0. This can be added to the labeled UTP.

• RNA polymerase 20 U / µL	**2 µL**

⇝ Polymerases (SP6, T3, or T7) are used for the synthesis of anti-sense and sense riboprobes. The latter serve as a control for the determination of the background.

• α^{35}S-UTP nucleotide	**50 µCi**
• Sterile water	**to 20 µL**

⇝ 1000 µCi / mmol

b. Mix and centrifuge.

⇝ All the components of the mixture must be at the bottom of the tube.

c. Incubate. **60 min at 37°C**

⇝ Do not prolong incubation time, to avoid degradation of the RNA.

2. Digestion of the DNA

a. Add:
• DNase 1 U / µL	**2 µL**
b. Incubate.	**15 min at 37°C**

⇝ Final concentration 2 units
⇝ To eliminate the DNA strands specifically (elimination of the template)

Typical Protocols

    c. Add:
- 100 mM EDTA   2 µL   ⇰ The enzyme is inhibited by EDTA (final concentration 4 mM)

3. Purification

a. Prepare the Sephadex G50 column.
b. Centrifuge.   **3000 g for 4 min**   ⇰ Elimination of the solvents
c. Add:
- Labeling medium   20 µL

d. Centrifuge.   **3000 g for 4 min**   ⇰ To recover the cRNA in an Eppendorf tube

4. Precipitation

a. Add 10 mg/mL tRNA.   2 µL   ⇰ Nucleic acid of large size (carrier), which facilitates precipitation

b. Precipitate with alcohol: ⇰ To eliminate free nucleotides
- 7.5 M ammonium acetate   9 µL   ⇰ 1:10 of final volume
- Ethanol 100% at –20°C   78 µL   ⇰ Two to three times the reactive volume to be precipitated

c. Incubate.   **30 to 60 min at – 80°C, or overnight at – 20°C**   ⇰ Precipitation

d. Centrifuge.   **≥ 14,000 g 30 min at 4°C**

e. Remove the supernatant.   ⇰ Orient the tube
f. Dry the pellet.

8.1.4 Antigenic cDNA Probe

⇰ Gloves must be worn.
⇰ Labeling by nick translation

Reactive medium

Precipitation

1. Reactive medium

a. To the dry pellet, add the following reagents in the indicated order:
- Nick buffer 5X   4 µL   ⇰ Some mixtures contain the buffer, the non-labeled nucleotides, and the enzymes.

• Nonlabeled dNTP	1 µL each	⇒ Concentration (0.5 mM dATP, dCTP, dGTP, 0.1 mM dTTP)
• Labeled dUTP	2 µL	⇒ Concentration 1 mM labeled dUTP
• Plasmid DNA	x µL	⇒ To obtain ≈ 1 µg probe
• Enzyme	5 µL	
• Sterile water	to 20 µL	
b. Vortex, centrifuge.	≈1000 g at 4°C	⇒ Recovery of all the reagents at the bottom of the tube
c. Incubate.	90 min to 3 h at 15°C	⇒ Incorporation of labeled nucleotides (after 2 h at 15°C all the missing bases are replaced)
d. Add:		⇒ **Optional**
• 10 mM EDTA	2 µL	⇒ To stop the reaction; also possible to stop it by heating (10 min at 65°C)
e. Vortex, centrifuge.		⇒ Mix, and remove 1 µL to measure total radioactivity before extraction of the probe

2. *Precipitation*

a. Add 10 mg/mL tRNA	2 µL	⇒ Nucleic acid of large size (carrier), which facilitates precipitation
b. Precipitate with alcohol:		⇒ To eliminate free nucleotides
• 7.5 M ammonium acetate	9 µL	⇒ 1:10 of final volume
• Ethanol 100% at –20°C	72 µL	⇒ Two to three times the reactive volume to be precipitated
c. Incubate.	30 to 60 min at –80°C or overnight at –20°C	⇒ Precipitation
d. Centrifuge.	≥ 14,000 g for 30 min at 4°C	
e. Remove the supernatant.		⇒ Orient the tube
f. Dry the pellet.		

Typical Protocols

8.1.5 Antigenic Oligonucleotide Probe

⇝ Gloves must be worn.
⇝ Labeling by 3' extension

Reactive medium
Precipitation

1. Reactive medium
a. Place the following reagents in a sterile Eppendorf tube in the indicated order:

• Oligonucleotides 10 to 100 pmol	**6 µL**
• Buffer 5X	**5 µL**
• 25 mM CoCl$_2$	**2.5 µL**
• Labeled nucleotide	**2.5 µL**
• TdT 25 units / µL	**1 µL**
• Sterile distilled water	**to 25 µL**

⇝ Concentration 10 to 50 pmol

⇝ Potassium cacodylate 1 M Tris – HCl; 125 mM; BSA 1.25 mg/mL

⇝ Digoxigenin–11 – dUTP (2.5 nmol / 25 µL)
⇝ Final concentration 1 U/µL

b. Incubate. **60 min at 37°C**
c. Stop the reaction:
  • 5 mM EDTA **100 µL**

⇝ Optional

2. Precipitation
a. Add:
  • 10 mg/mL tRNA **2 µL**
b. Precipitate with alcohol:
  • 7.5 M ammonium acetate **10 µL**
  • Ethanol 100% at –20°C **80 µL**

⇝ To eliminate free nucleotides
⇝ 1:10 of final volume
⇝ Two to three times the reactive volume to be precipitated

c. Incubate. **30 to 60 min at –80°C, or overnight at –20°C**

⇝ Precipitation

d. Centrifuge. **≥ 14,000 g for 30 min at 4°C**
e. Remove the supernatant.
f. Dry the pellet.

⇝ Orient the tube

8.1.6 Antigenic cRNA Probe

⇝ Gloves must be worn.
⇝ Labeling by *in vitro* transcription

In vitro *transcription*
Digestion of DNA
Purification
Precipitation

1. In vitro *transcription*
a. Place the following reagents in an Eppendorf tube, in ice, in the indicated order:

• Linearized plasmid	**1 µg**
• Transcription buffer 5X	**2 µL**
• 20 U/µL RNasin	**1 µL**

⇝ Final concentration 1 U/µL

8.1 Preparation of Probes

• rNTP mix:	2 µL	⇝ RNase inhibitor, to avoid the degradation of the synthesized RNA. This reagent is not resistant to high temperatures. It is added at the last moment.
— 10 mM ATP		⇝ Final concentration 0.4 mM
— 10 mM GTP		⇝ First, prepare a mixture of all the components, but without DNA and RNA polymerase, neutralized at pH 7.0. It can be added to the labeled UTP.
— 10 mM CTP		
— 6.5 mM UTP		
• RNA polymerase 20 U/µL	2 µL	⇝ Polymerases (SP6, T3, or T7) are used for the synthesis of anti-sense and sense ribo-probes. The latter serve as a control for the determination of the background noise.
• Labeled nucleotide	2 µL	⇝ (Biotin, digoxigenin, fluorescein) UTP (3.5 mM)
• Sterile water	to 20 µL	
b. Vortex and centrifuge.		⇝ Necessary for all the reagents to be at the bottom of the tube
c. Incubate.	2 h at 37°C	⇝ Do not prolong incubation time, to avoid degradation of the RNA.

2. Digestion of DNA

a. Add:
- 1 U/µL DNase   2 µL
- ⇝ Optional

b. Incubate.   15 min at 37°C   ⇝ To eliminate the DNA strands specifically (elimination of the template)

c. Add:
- 100 mM EDTA   2 µL   ⇝ Enzyme inhibited by EDTA (final concentration 4 mM)

3. Purification

a. Prepare the Sephadex G50 column.

b. Centrifuge.   3000 g for 4 min   ⇝ Elimination of the solvents

c. Add:
- Labeling medium   22 µL

d. Centrifuge.   3000 g for 4 min   ⇝ Recovery of the cRNA in an Eppendorf tube

4. Precipitation

a. Add 10 mg/mL tRNA.   2 µL   ⇝ Nucleic acid of large size (carrier), which facilitates precipitation

b. Precipitate with alcohol:
- 7.5 M ammonium acetate   9 µL   ⇝ To eliminate free nucleotides; ⇝ 1:10 of final volume
- Ethanol 100% at –20°C   72 µL   ⇝ Two to three times the reactive volume to be precipitated

c. Incubate.   30 to 60 min, – 80°C, or overnight, – 20°C   ⇝ Precipitation

d. Centrifuge.   ≥ 14,000 g for 30 min at 4°C

e. Remove the supernatant.   ⇝ Orient the tube

f. Dry the pellet

8.2 PREPARATION OF TISSUE ⇝ *See* Chapter 2.

8.2.1 Frozen Tissue

Fixation
Cryoprotection
Congelation
Storage
Production of frozen sections
Drying
Fixation
Dehydration
Drying
Storage

1. **Fixation**
   - 4% PF in 100 mM phosphate buffer — 3 to 24 h at 4°C
   - Buffer rinsing — 4 × 15 min

   ⇝ *See* Section 2.2.4.
   ⇝ *See* Appendix B4.4.2.
   ⇝ *See* Appendix B3.3

2. **Cryoprotection**
   - 30 to 50% saccharose — 2 to 24 h at 4°C

   ⇝ *See* Section 2.3.3
   ⇝ *See* Appendix B2.19.
   ⇝ Cryoprotection is total when the tissue is at the bottom of the receptacle.

3. **Freezing**
   ⇝ *See* Section 2.3.4.

4. **Storage**
   ⇝ Storage is possible.

5. **Production of frozen sections** — on pretreated slides — 10 µm
   ⇝ *See* Section 2.3.5.
   ⇝ *See* Appendix A3

6. **Drying**

7. **Fixation**
   - 4% PF in 100 mM phosphate buffer — 10 to 20 min at RT
   - Buffer rinsing — 2 × 5 min

   ⇝ *See* Section 2.2.4.
   ⇝ Use to facilitate the adhesion of the section to the slide.

8. **Dehydration**
   - 150 mM NaCl — 5 min
   - Alcohol 50%, 70%, 95% — 3 min / bath
   - Alcohol 100% — 2 × 3 min

9. **Drying**
   - In a vacuum jar — 60 min

   ⇝ If possible

10. **Storage** — −20°C
    ⇝ Storage is possible (hermetic boxes + desiccant)

Following steps:

- Protocols for radioactive probes ⇝ *See* Sections 8.3 and 8.4.
- Protocols for antigenic probes ⇝ *See* Sections 8.3 and 8.6.

8.2.2 Paraffin-Embedded Tissue

Fixation
Dehydration
Embedding
Inclusion
Microtomy
Drying
Dewaxing
Rehydration

1. **Fixation**
   - 4% PF in 100 mM phosphate buffer   **3 to 24 h at 4°C**   ↬ *See* Section 2.2.4.
   - Buffer rinsing   **4 × 15 min**   ↬ *See* Appendix B4.4.
   - 150 mM NaCl   **3 min at 4°C**   ↬ *See* Appendix B3.3.

2. **Dehydration**   ↬ *See* Section 2.4.3.
   - Alcohol 70%, 95%   **30 min / bath**
   - Alcohol 100%   **2 × 1 h**

3. **Embedding**
   - Xylene, or a substitute   **2 × 15 min**

4. **Inclusion**
   - Liquid paraffin   **56 to 58°C**
   — Tissue < 3 mm   **2 × 90 min**
   — Tissue > 3 mm   **Overnight**
   - Cooling the paraffin   **4°C**   ↬ Storage is possible.

5. **Microtomy**
   - Sections on pretreated slides   **5 to 20 μm**   ↬ *See* Section 2.4.4.

6. **Drying**   **Overnight at 56°C**   ↬ It is possible of store the sections (hermetic boxes + desiccant).

7. **Dewaxing**   ↬ **Dewaxing is an indispensable step**, to be carried out extemporaneously (*see* Section 2.4.5).
   - Xylene, or a substitute   **2 × 5 min**

8. **Rehydration**
   - Alcohol 100%, 95%, 70%   **3 min/bath**
   - 150 mM NaCl   **3 min**   ↬ **Carry out the pretreatments immediately.**

Following steps:
- Protocols for radioactive probes   ↬ *See* Sections 8.3 and 8.4.
- Protocols for antigenic probes   ↬ *See* Sections 8.3 and 8.6.

8.2.3 Preparation of Cells

⇝ *See* Section 2.4.

8.2.3.1 In suspension

1. *Depositing the suspension* ⇝ *See* Section 2.2.4.2.
   - Cytocentrifugation
   - Smears
   - Freezing in pellet form
   - Embedding of pellets in paraffin

8.2.3.2 In culture

⇝ *See* Section 2.2.4.3.

1. *Detaching the cells*
   - Mechanical   5 to 15 min
   - Enzymatic   at 37°C
   - Buffer rinsing   3 × 5 min   ⇝ Or culture medium, without serum
          at 4°C

2. *Depositing the cell suspension*   ⇝ *See* Section 2.4.3.
   - Smears
   - Direct deposition
   - Cytocentrifugation   100 µL
   - Freezing of pellets   ⇝ *See* Section 8.2.1.
   - Inclusion of pellets in paraffin   ⇝ *See* Section 8.2.2.

3. *Drying*

8.2.3.3 Fixation

1. *Fixation*
   - 4% PF in 100 mM phosphate buffer   10 min, at 4°C   ⇝ *See* Appendix B4.4.2.
   - Buffer rinsing   3 × 5 min at 4°C   ⇝ *See* Appendix B3.3.

2. *Dehydration*
   - Alcohol 50%, 70%, 90%, 100%   3 min / bath

3. *Drying*
   - In a vacuum jar   1.5 to 2 h at RT

4. *Storage*   –20°C   ⇝ Possibility of storage (hermetic boxes + desiccant)

Following steps:

- Protocols for radioactive probes   ⇝ *See* Section 8.3.
- Protocols for antigenic probes   ⇝ *See* Section 8.3.

8.3 PRETREATMENTS

↪ *See* Chapter 3.
↪ **Gloves must be worn.**

1. Rehydration

• 150 mM NaCl	**3 min**
• 100 mM phosphate buffer	**3 min**

↪ **If sections are stored at –20°C:** leave the boxes containing the sections for at least 2 h at room temperature before opening.

↪ *See* Appendix B3.3.1.

2. Fixation

• 4% PF in 100 mM phosphate buffer	**15 min at RT**
• Buffer rinsing	**2 × 5 min**

↪ *See* Section 3.4.2.
↪ *See* Appendix B4.4.2.

3. Deproteinization — Incubate successively in the following solutions:

• 20 mM Tris buffer; 2 mM CaCl$_2$; pH 7.6	**5 min**
• Proteinase K in Tris – HCl/CaCl$_2$ buffer	**15 min at 37°C**

↪ *See* Section 3.6.2.1.2.
↪ *See* Appendix B3.6.2.
↪ Change buffer, rinse.
↪ *See* Appendix B2.17.
↪ Concentration 1 to 3 µg/mL for frozen sections
↪ Concentration 5 to 10 µg/mL for paraffin sections

• 20 mM Tris buffer; 2 mM CaCl$_2$; pH 7.6	**5 min**
• 100 mM phosphate buffer	**5 min**

4. Fixation

• 4% PF in buffer	**5 min**
• Buffer rinsing	**5 min**
• 150 mM NaCl	**2 min**

↪ **Indispensable**
↪ 100 mM phosphate buffer *(see* Appendix B4.4)

5. Acetylation

• Triethanolamine buffer	**5 min**
• Add: acetic anhydride	**1 mL /200 mL**
	1 min
• 100 mM phosphate buffer	**5 min**
• 150 mM NaCl	**2 min**

↪ **Optional.** *See* Section 3.7.
↪ *See* Appendix B3.1. Shake the sections.
↪ Final concentration 0.5%, with **constant shaking**

6. Dehydration

• Alcohol 70%, 95%, 100%	**2 min / bath**

↪ *See* Section 3.8.2.

7. Drying

• In a vacuum jar	**30 to 60 min**

↪ Storage is possible at –20°C.

Typical Protocols

  • In air   60 min

8. Prehybridization   ↪ **Optional**. *See* Section 3.9.2

  • Hybridization buffer   ≈100 µL/section
  • Incubation time   1 to 2 h
              at RT

9. Denaturation   ↪ **Optional**. Only with double-stranded DNA
                  (*See* Section 3.10.2)

 a. Denature target DNA
  • By heat   5 min   ↪ Alternative: by microwave heating (750 W),
              at 94°C 3×3 min
 b. Denature probe DNA
  • By heat   3 min
              at 100°C
 c. Plunge the probe immediately into ice.
 d. Place the slides immediately on ice.

Following step: Hybridization   ↪ *See* Section 8.4 or Section 8.6.

8.4 RADIOACTIVE PROBES: HYBRIDIZATION, WASHING

8.4.1 Radioactive cDNA Probe
↪ *See* Section 1.2.1.
↪ $\alpha^{35}S$ or ^{33}P

Preparation of the hybridization buffer
Preparation of the reactive medium
Hybridization
Washing
Dehydration

1. Preparation of the hybridization buffer   ↪ Stock solution (500 µL)

 a. Place the following reagents in a sterile
 Eppendorf tube, in the indicated order:
  • 50% dextran sulfate   100 µL   ↪ *See* Appendix 2.22
                     ↪ Final concentration 10%
  • SSC 20X   100 µL   ↪ Final concentration 4X
  • 100% formamide   250 µL   ↪ Final concentration 50%
  • Denhardt's solution 50X   10 µL   ↪ Final concentration 1X
  • tRNA 10 mg / mL   12.5 µL   ↪ Final concentration 250 µg/mL
  • DNA 10 mg / mL   12.5 µL   ↪ Final concentration 250 µg/mL
  • Sterile water   to 500 µL
 b. Vortex.
 c. Centrifuge.   ↪ Elimination of bubbles

Storage — Possibility of storage at −20°C ⇒ Without probe

2. Preparation of the reactive medium

⇒ Hybridization medium + probe

a. Add to the hybridization buffer:
- Labeled cDNA probe   < 1 µg   ⇒ 0.1 to 0.2 µg/mL, according to the specific activity of the probe
- Hybridization buffer   to 1 mL

b. Mix on vortex, centrifuge.
c. Denature.   6 min at 100°C
d. Cool rapidly in ice.
e. Add:
- 10 mM DTT   5 µL   ⇒ **Indispensable** in cases of probes labeled with ^{35}S
⇒ At the moment of use

Storage — Possibility of storage at −20°C for some days   ⇒ Phenomenon of radiolysis

3. Hybridization

a. Immediately place the reactive medium on the sections.
- Reactive medium   20 µL/section   ⇒ For a 22 × 22 mm coverslip
                            50 µL/section   ⇒ For a 24 × 60 mm coverslip

b. Put the coverslip in place.
c. Seal with rubber cement.   ⇒ **Optional**
d. Incubate.   Overnight at 40°C
⇒ Possible to reduce time
⇒ In the presence of SSC 5X
⇒ In moisture chamber

4. Washing   RT

- SSC 5X   5 min   ⇒ **Probe is radioactive** (use appropriate means of disposal)

- SSC 2X + 50% formamide   1 h
- SSC 2X   2 × 30 min
- SSC 1X   1 h
- SSC 0.5X   1 h at 50 to 55°C   ⇒ Optional step
- SSC 0.5X   30 min   ⇒ Optional step
- SSC 0.1X   30 to 60 min   ⇒ Optional step

5. Dehydration

- Alcohol 70%, 95%   2 min/bath
- Alcohol 100%   2 × 2 min

Drying

Following step: Autoradiography   ⇒ *See* Section 8.5.

8.4.2 Radioactive Oligonucleotide Probe

⇝ *See* Chapter 1.
⇝ α(^{35}S or ^{33}P)–NTP

Preparation of the hybridization buffer
Preparation of the reactive medium
Hybridization
Washing
Dehydration

1. Preparation of the hybridization buffer

⇝ Stock solution (500 µL)

a. Place the following reagents in a sterile Eppendorf tube, in the indicated order:
• 50% dextran sulfate   100 µL

⇝ *See* Appendix B2.22
⇝ Because of the difficulty of pipetting this component, the other components are added to it.
⇝ Final concentration 10%

• SSC 20X   100 µL
• 100% formamide   250 µL
• Denhardt's solution 50X   10 µL
• tRNA 10 mg / mL   12.5 µL
• DNA 10 mg / mL   12.5 µL
• Sterile water   to 500 µL

⇝ Final concentration 4X
⇝ Final concentration 50%
⇝ Final concentration 1X
⇝ Final concentration 250 µg / mL
⇝ Final concentration 250 µg / mL

b. Mix on vortex.
c. Centrifuge.

⇝ Elimination of bubbles

Storage — Possibility of storage at – 20°C

⇝ Without probe

2. Preparation of the reactive medium

⇝ Hybridization medium + probe

a. To the hybridization buffer, add:
• Oligonucleotide probe   2.5 pmol

⇝ 1 to 5 pmol / mL, according to the specific activity of the probe

• Hybridization buffer   to 1 mL
b. Mix on vortex, centrifuge.

Storage — Possibility of storage at – 20°C for some days

⇝ Phenomenon of radiolysis

3. Hybridization

a. Immediately place the reactive medium on the sections:
• Reactive medium   20 µL/section
                           50 µL/section

⇝ For a 22 × 22 mm coverslip
⇝ For a 24 × 60 mm coverslip

b. Cover with a coverslip.
c. Seal with rubber cement.

⇝ Optional

d. Incubate.	**Overnight at 40°C**	↝ Possible to reduce the time ↝ In the presence of SSC 5X ↝ Moisture chamber
4. Washing • SSC 5X	**RT** **5 min**	↝ **Probe is radioactive** (use appropriate means of disposal)
• SSC 2X + 50% formamide	**1 h**	
• SSC 2X	**1 h**	↝ Temperature to be determined
• SSC 1X	**1 h**	
• SSC 0.5X	**1 h**	↝ **Optional step**
5. Dehydration		
• Alcohol 70%	**2 min**	
• Alcohol 95%	**2 min**	
• Alcohol 100%	**2 × 2 min**	
Drying	**1 to 3 h**	

Following step — Autoradiography ↝ *See* Section 8.5

8.4.3 Radioactive cRNA Probe

↝ *See* Chapter 1
↝ α(^{35}S or ^{33}P)–NTP

Preparation of the hybridization buffer
Preparation of the reactive medium
Hybridization
Washing
Dehydration

1. Preparation of the hybridization buffer

↝ Possibility of storage at –20°C

a. Place the following reagents in a sterile Eppendorf tube, in the indicated order:

• 50% dextran sulfate	**100 µL**	↝ *See* Appendix B2.22 ↝ Because of the difficulty of pipetting this component, the other components are added to it. ↝ Final concentration 10%
• SSC 20X	**100 µL**	↝ Final concentration 4X
• 100% formamide	**250 µL**	↝ Final concentration 50%
• Denhardt's solution 50X	**10 µL**	↝ Final concentration 1X
• tRNA 10 mg / mL	**12.5 µL**	↝ Final concentration 250 µg/mL
• Sterile water	**to 500 µL**	
b. Vortex.		
c. Centrifuge.		↝ Elimination of bubbles

Storage — Possibility of storage at –20°C

2. Preparation of the reactive medium

a. Add to the hybridization buffer:

• Labeled cRNA probe	**0.01 to 0.1 µg / mL**
• Hybridization buffer	**to 1 mL**

b. Mix on vortex, centrifuge.
c. Add:

• 10 m*M* DTT	**5 µL**

↪ Hybridization medium + probe

3. Hybridization

a. Immediately place the reactive medium on the sections:

• Reactive medium	**20 µL / section**	↪ For a 22 × 22 mm coverslip
	50 µL / section	↪ For a 24 × 60 mm coverslip

b. Cover with a coverslip.
c. Seal with rubber cement.
d. Incubate. **Overnight**
45 to 50°C

↪ **Optional**
↪ Can reach 60°C
↪ SSC 5X
↪ Moisture chamber

4. Washing

• SSC 5X	**5 min**	↪ Room temperature
• SSC 2X + 50% formamide	**30 min**	↪ Room temperature
• SSC 2X	**30 min** **55°C**	
• SSC 1X	**30 min**	↪ Room temperature
• SSC 0.5X	**30 min**	↪ Room temperature
• SSC 0.1X	**30 min**	↪ Room temperature
• SSC 2X + RNase 10 µg/mL	**30 to 60 min** **37°C**	↪ Water bath, with shaking ↪ Stock solution (RNase 10 mg / mL) (*see* Appendix B2.18)

↪ This step also involves the inhibition of RNases.

5. Dehydration

• Alcohol 70%, 95%	**1 min / bath**
• Alcohol 100%	**2 × 1 min**

Drying

Following step — Autoradiography

↪ *See* Section 8.5

8.5 AUTORADIOGRAPHY

8.5.1 Macroautoradiography

↪ *See* Section 5.1.3.5.

1. Place film.

↪ **Under safelight**
↪ Store in autoradiographic boxes

2. Expose. **1 to 3 days at 4°C**

3. Develop.

8.5.2 Microautoradiography

↪ *See* Section 5.1.4.

1. Dilute the emulsion:
- LM1   **2:1**
- NTB2   **1:2**

↪ **Under safelight**
↪ Emulsion/distilled water 2:1 ($^v/_v$)
↪ Emulsion/distilled water 1:2 ($^v/_v$)

2. Liquefy the emulsion. **1 h at 43°C**

↪ Necessary for the emulsion to be perfectly homogeneous

3. Dip the slides, and withdraw them vertically with a uniform motion.

4. Dry. **2 h to overnight at RT**

5. Expose. **A few days at 4°C**

↪ Hermetic boxes + desiccant

8.5.3 Autoradiographic Revelation

↪ **Under safelight**

- Developer 1:1   **4 min at 17°C**
- Rinse in distilled water   **30 s**
- 30% sodium thiosulfate   **5 min**
- Rinse in running water   **30 min**
- Rinse in distilled water   **1 min**

↪ *See* Appendix B5.1

↪ *See* Appendix B5.2

8.5.4 Counterstaining

↪ **Optional**

- Toluidine blue   **1 min**   ↪ *See* Appendix B7.1.
- Cresyl violet   **30 s**   ↪ *See* Appendix B7.2.
- Hematoxylin–eosin   **30 s + 30 s**   ↪ *See* Appendix B7.4 / B7.3.
- Harris's hematoxylin **30 s to 1 min** ↪ *See* Appendix B7.4.
- Methyl green   **3 min**   ↪ *See* Appendix B7.6.

8.5.5 Mounting after Dehydration

↪ *See* Appendix B8.2.

1. **Rinse in distilled water** 1 min

2. **Dehydrate:** ↪ *See* Section 6.2.1.

   - Alcohol 70%, 95%, 100% **3 × 5 min / inhibitor**
   - Alcohol **3 × 5 min**
   - Xylene **2 × 5 min** ↪ It is possible to prolong the time of use of these two latter inhibitors without risk.

3. **Mount** — Use the mounting medium on a slide humidified with xylene for better spreading. Put the coverslip in place immediately. ↪ Examples of mounting medium are Permount, Eukitt, or Depex. Use a heating block for better spreading.

8.5.6 Modes of Observation

↪ *See* Section 6.3.

- Bright field ↪ The silver grains appear black.
- Dark field ↪ Do not counterstain.
- Epipolarization ↪ The silver grains appear bright.

8.6 ANTIGENIC PROBES: HYBRIDIZATION, WASHING

8.6.1 Antigenic cDNA Probe

⇒ *See* Section 1.2.2.
⇒ (Biotin, digoxigenin, fluorescein)–dUTP

Preparation of the hybridization buffer
Preparation of the reactive medium
Hybridization
Washing

1. Preparation of the hybridization buffer

a. Put the following reagents in a sterile Eppendorf tube, in the indicated order:

• 50% dextran sulfate	**100 µL**

⇒ *See* Appendix B2.22.
⇒ **Optional**
⇒ Because of the difficulty of pipetting this component, the other components are added to it.
⇒ Final concentration 10%

• 100% formamide	**250 µL**
• SSC 20X	**100 µL**
• tRNA 10 mg / mL	**12.5 µL**
• DNA 10 mg / mL	**12.5 µL**
• Denhardt's solution 50X	**10 µL**
• Sterile water	**to 500 µL**

⇒ Final concentration 50%
⇒ Final concentration 4X
⇒ Final concentration 250 µg / mL
⇒ Final concentration 250 µg / mL
⇒ Final concentration 1X

b. Vortex.

c. Centrifuge.

⇒ Elimination of bubbles

Storage — Possibility of storage at –20°C

2. Preparation of the reactive medium

⇒ Hybridization buffer + probe

a. Add to the hybridization buffer:

• Labeled cDNA probe	**3 to 10 µg**
• Hybridization buffer	**to 1 mL**

b. Mix on a vortex, centrifuge.

c. Denature. **6 min at 100°C**

⇒ **Indispensable**

d. Cool rapidly in ice.

Typical Protocols

3. Hybridization

a. Immediately place the reactive medium on the sections:

• Reactive medium	20 µL / section	↪ For a 22 × 22 mm coverslip
	50 µL / section	↪ For a 24 × 60 mm coverslip

b. Put the coverslip in place.
c. Seal with rubber cement.
d. Incubate. **Overnight at 40 to 42°C**   ↪ Moisture chamber
↪ SSC 5X

4. Washing

↪ Room temperature

• SSC 4X	5 min
• SSC 2X	2 × 30 min
• SSC 1X	60 min

Following step — Immunocytological detection   ↪ *See* Section 8.7.

8.6.2 Antigenic Oligonucleotide Probe

↪ *See* Chapter 1.
↪ (Biotin, digoxigenin, fluorescein)–dUTP

Preparation of the hybridization buffer
Preparation of the reactive medium
Hybridization
Washing

1. Preparation of the hybridization buffer

↪ Possibility of storage at –20°C

a. Place the following reagents in a sterile Eppendorf tube, in the indicated order:

• 50% dextran sulfate	100 µL	↪ *See* Appendix B2.22.
		↪ Because of the difficulty of pipetting this component, the other components are added to it.
		↪ Final concentration 10%
• SSC 20X	100 µL	↪ Final concentration 4X
• 100% formamide	250 µL	↪ Final concentration 50%
• Denhardt's solution 50X	10 µL	↪ Final concentration 1X
• tRNA 10 mg / mL	12.5 µL	↪ Final concentration 250 µg / mL
• DNA 10 mg / mL	12.5 µL	↪ Final concentration 250 µg / mL
• Sterile water	to 500 µL	

b. Vortex.
c. Centrifuge.   ↪ Elimination of bubbles

2. Preparation of the reactive medium

a. Add the hybridization buffer to the labeled probe:
- Labeled oligonucleotide probe   **10 to 100 pmol**
- Hybridization buffer   **to 500 µL**

b. Mix on vortex, centrifuge.

3. Hybridization

a. Immediately place on the dried sections:
- Reactive medium   **20 µL / section**
    50 µL / section

b. Put the coverslip in place.
c. Seal with rubber cement.
d. Incubate.   **Overnight at 40°C**

4. Washing

- SSC 5X   **5 min**
- SSC 2X   **60 min**
- SSC 1X   **60 min**
- SSC 0.5X   **60 min**

Following step: Immunocytological detection.

8.6.3 Antigenic cRNA Probe

Preparation of the hybridization buffer
Preparation of the reactive medium
Hybridization
Washing

1. Preparation of the hybridization buffer

a. Place the following reagents in a sterile Eppendorf tube, in the indicated order:
- SSC 20X   **100 µL**
- 100% formamide   **250 µL**
- Denhardt's solution 50X   **50 µL**
- tRNA 10 mg / mL   **12.5 µL**
- Sterile water   **to 500 µL**

b. Vortex.
c. Centrifuge.

↪ Hybridization buffer + probe

↪ Final concentration 20 to 200 pmol/mL

↪ To dissolve the pellet fully

↪ For a 22 × 22 mm coverslip
↪ For a 24 × 60 mm coverslip

↪ Optional

↪ SSC 5X
↪ Moisture chamber

↪ Room temperature

↪ See Section 8.7.

↪ *Gloves must be worn.*
↪ (Biotin, digoxigenin, fluorescein) – UTP

↪ Possibility of storage at –20°C

↪ Final concentration 4X
↪ Final concentration 50%
↪ Final concentration 5X
↪ Final concentration 250 µg / mL

↪ Elimination of bubbles
↪ Possibility of storage at –20°C

2. Preparation of the reactive medium ↪ Hybridization buffer + probe

a. Add the labeled probe to the hybridization buffer:

- Labeled probe **X µL** ↪ Final concentration 0.2 to 2 µg/mL

b. Vortex, centrifuge. ↪ To dissolve the pellet fully

3. Hybridization

a. Immediately place the reactive medium on the dried sections:

- Reactive medium **20 µL/section** ↪ For a 22 × 22 mm coverslip
  50 µL/section ↪ For a 24 × 60 mm coverslip

b. Put the coverslip in place.

c. Seal with rubber cement. ↪ **Optional**

d. Incubate. **Overnight at 50°C** ↪ SSC 5X
  ↪ Moisture chamber

4. Washing

- SSC 1X **15 min** ↪ Room temperature

- SSC 0.25X **2 × 10 min at 50°C** ↪ Water bath, with shaking

- SSC 2X **2 min at RT**

- SSC 2X + RNase 10 µg/mL **30 to 60 min at 37°C** ↪ Water bath, with shaking
  ↪ Stock solution (RNase 10 mg/mL) (*see* Appendix B2.18)

- SSC 2X **2 × 5 min** ↪ Room temperature

Following step: Immunocytological detection. ↪ *See* Section 8.7.

8.7 IMMUNOCYTOLOGICAL DETECTION

8.7.1 Immunohistological Reaction

1. **Rinsing**
   - Buffer: 50 mM Tris –
     HCl; 300 mM NaCl   5 min   ↬ *See* Appendix B3.6.

2. **Saturate nonspecific sites**
   - Blocking buffer   15 to 30 min   ↬ *See* Appendix B6.1.1
     (50 mM Tris–HCl;   ↬ Possible to replace goat serum with another
     300 mM NaCl;   blocking agent
     1% goat serum)   ↬ With the addition, or not, of Triton X-100
     ($\pm 0.1\%$)

3. **Deposit** the conjugated anti-hapten antibody,   ↬ IgG, Fab fragments coupled to an enzyme
   diluted in blocking buffer: **(alkaline phosphatase or peroxidase)**
   - Diluted 1:500   100 µL/section   ↬ Dilution possible up to 1 / 5000

4. **Incubate.**   2 h to overnight
                    at RT

5. **Rinse** with 50 mM Tris– 3×10 min
   HCl buffer; 300 mM NaCl

Following step — Revelation of the enzyme.   ↬ Alkaline phosphatase (*see* Section 7.2)
    ↬ Peroxidase (*see* Section 7.3)

8.7.2 Revelation with Alkaline Phosphatase

↬ Inhibition of endogenous phosphatase activity (*see* Appendix B6.1.2)

1. **Change of buffer:**   ↬ Basic buffer
   - 50 mM Tris–HCl buffer;   2 to 10 min   ↬ *See* Appendix B3.6.6.
     300 mM NaCl; 50 mM
     MgCl$_2$; pH 9.5

2. **Prepare the revelation solution:**   ↬ Substrates (NBT – BCIP or Fast – Red)
   a. NBT – BCIP   ↬ Prepare at time of use (*see* Appendix
                B6.2.1.1)
   - NBT (75 mg/mL)   45 µL   ↬ Ready-to-use solutions and tablets available
   - BCIP (50 mg/mL)   35 µL
   - 50 mM Tris–HCl buffer;   10 mL   ↬ *See* Appendix B3.6.6.
     300 mM NaCl; 50 mM
     MgCl$_2$; pH 9.5

Typical Protocols

    b. Fast – Red
- Naphthol substrate   **2 mg**
- Dimethylformamide   **200 μL**
- 50 mM Tris–HCl buffer;   **9.8 mL**   ↪ *See* Appendix B3.6.6.
  300 mM NaCl; 50 mM
  MgCl$_2$; pH 9.5
- Fast–Red 1:250   **10 mg**   ↪ MW = 273.20.

3. **Place the substrate on the sections:**

- Revelation solution   **100 μL / section**

4. **Incubate** under visual surveillance.   **1 to 24 h, at RT in darkness**   ↪ Use moisture chamber (buffer).

5. **Stop the reaction** with distilled water.   **1 min**

Following steps:

- Counterstaining   ↪ **Optional.** *See* Section 7.7.4.
- Mounting   ↪ *See* Section 7.7.5.

8.7.3 Revelation with Peroxidase

↪ Inhibition of endogenous peroxidase activity (*see* Appendix B6.2.2)

1. Prepare the revelation solution.   ↪ Substrates (DAB, 4-chloro-1-naphthol, or AEC)

    a. DAB   ↪ Prepare at time of use
     ↪ *See* Appendix B6.2.2.1.

- DAB or one tablet   **10 mg**
- Revelation buffer   **10 mL**   ↪ 50 mM Tris–HCl buffer; 300 mM NaCl; pH 7.6 (*see* Appendix B3.6.5)

Filter, and add to the filtrate:
- 30% hydrogen peroxide   **1 μL**

    b. 4-Chloro-1-naphthol   ↪ Prepare at time of use
     ↪ *See* Appendix B6.2.2.2.

- 4-Chloro-1-naphthol   **50 mg / 230 μL ethanol 95%**
- Revelation buffer   **50 mL**   ↪ Tris–NaCl buffer; pH 7.6 (*see* Appendix B3.6.5)

Filter, and add to the filtrate:
- 30% hydrogen peroxide   **5 μL**

    c. 3-Amino-9-ethyl carbazole (AEC)   ↪ *See* Appendix B6.2.2.3.
- AEC   **4 mg**
- Dimethylformamide   **1 mL**

• Revelation buffer	14 mL	↪ 100 mM sodium acetate buffer; pH 5.2

Filter, and add to the filtrate:
- 30% hydrogen peroxide. **1.5 µL**

2. Place the substrate on the sections. 100 µL/section

3. Incubate under visual surveillance. 3 to 10 min at RT

4. Stop the reaction with distilled water. 1 min

Following steps:

• Counterstaining		↪ **Optional.** *See* Section 8.7.4.
• Mounting		↪ *See* Section 8.7.5.

8.7.4 Counterstaining

↪ **Optional** (*see* Appendix B7)

1. Counterstaining

• 0.02% toluidine blue	**1 to 10 min**	↪ *See* Appendix B7.1.
• Cresyl violet	**1 to 5 min**	↪ *See* Appendix B7.2.
• Hematoxylin – eosin	**30 s/stain**	↪ *See* Appendix B7.4 (hematoxylin), and Appendix B7.3 (eosin). Wash in water between stains.
• Harris's hematoxylin	**30 s to 1 min**	↪ *See* Appendix B7.4 .
• Rapid nuclear red	**1 min**	↪ *See* Appendix B7.5.
• Methyl green	**1 to 3 min**	↪ *See* Appendix B7.6.

2. Washing
- Running water **1 min**

Following step: Mounting.

↪ **Optional** (*see* Section 8.7.5)

8.7.5 Mounting

↪ *See* Appendix B8.

1. Mounting in an aqueous medium

↪ *See* Appendix B8.1.
↪ If NBT – BCIP, Fast – Red, 4-chloro-1-naphthol, AEC
↪ If DAB, dehydrate (possible if NBT – BCIP)
↪ Possible to use commercial mounts

- PBS – glycergel **vol:vol**

2. Permanent mounting
   a. Dehydration

↪ *See* Appendix B8.2.
↪ *See* Section 6.2.1.

- Alcohol 70% **3 × 5 min**
- Alcohol 95% **3 × 5 min**

- Alcohol 100%   **3 × 5 min**
- Xylene   **2 × 5 min**

b. Mounting — Use the mounting medium on slides humidified with xylene for better spreading. Put the coverslip in place immediately.

3. **Observation**
   - Bright field

↪ It is possible to prolong the time of use of the two latter washings without risk.
↪ Examples of mounting medium are Permount, Eukitt or Depex. Use a heating block for better spreading.

↪ *See* Section 6.3.2

Examples of Observations

Examples of Observations

Method:
Macroautoradiography

Tissue:
Rat pituitary

Probe:
Oligonucleotide (30 mers)

Revelation:
Autoradiographic film (exposure 8 days)

Preparation:
Frozen sections

Labeling:
3' extension by addition of [^{35}S] dATP

Mode of observation:
Macroscopy

Signal:
Density of the autoradiographic film. The level of gray is proportional to that of the radioactive emission.

Results:

(A): Revelation of mRNA coding for the growth hormone (GH). The signal is present in the anterior lobe (AL) of the pituitary. The intermediate and posterior lobes are negative (*). The level of background noise is relatively high.

(B): Revelation of mRNA coding for the adrenocorticotrophic hormone (ACTH). The signal is present mainly in the intermediate lobe (IL) of the pituitary. The density of the signal is lower in the anterior lobe. The nervous lobe (*) is negative.

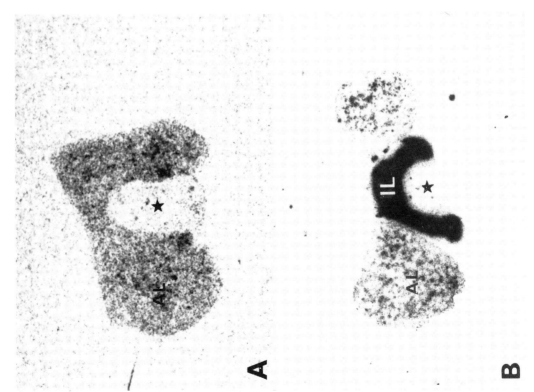

Examples of Observations

Method:
Microautoradiography

Tissue:
Rat pituitary

Probe:
cDNA (800 bp)

Revelation:
Autoradiographic emulsion (exposure 21 days)

Signal:
Black grains on a background staining (eosin)

Preparation:
Paraffin-embedded sections

Labeling:
Random priming incorporation of dNTP-[^{3}H]

Mode of observation:
Bright-field microscopy

Results:
Revelation of mRNA coding for growth hormone (GH)

(A): The signal is present in the anterior lobe (AL) of the pituitary. The cells of the intermediate lobe (IL) are negative. Background noise is nonexistent.

(B): All the cells in the anterior lobe are not positive. The signal is detectable mainly in the cytoplasm of somatotropic cells.

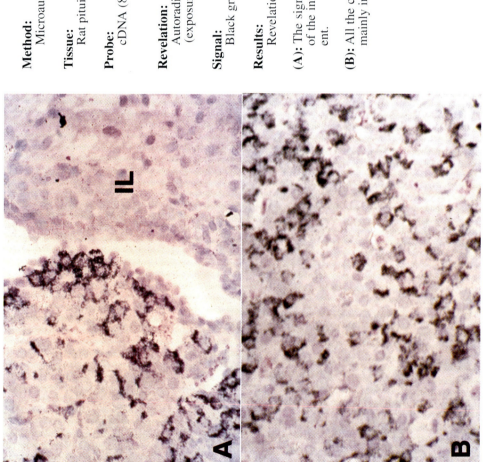

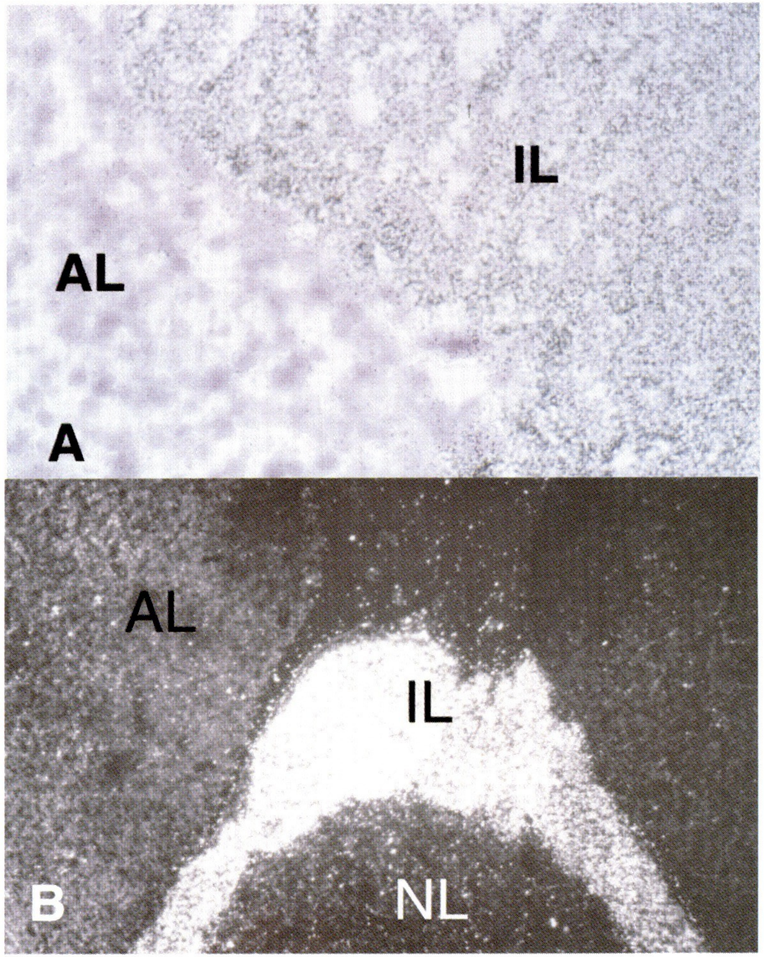

Examples of Observations

Method:
Microautoradiography

Tissue:
Rat pituitary

Probe:
Oligonucleotide (30 mers)

Preparation:
Frozen sections

Labeling:
3' extension by incorporation of [^{35}S] dATP

Revelation:
Autoradiographic emulsion (exposure 7 days)

Mode of observation:
(A) Bright-field microscopy
(B) dark-field microscopy

Signal:

(A): Black grains on a background staining (toluidine blue)

(B): Bright grains

Results: Revelation of mRNA coding for the adrenocorticotropic hormone (ACTH)

(A): The signal is present over the intermediate lobe (IL) of the pituitary. Over the anterior lobe (AL), it is difficult to visualize positive cells.

(B): The signal is present mainly in the intermediate lobe (IL) of the pituitary. The density of the signal is weaker in the anterior lobe (AL). The posterior lobe (PL) is negative.

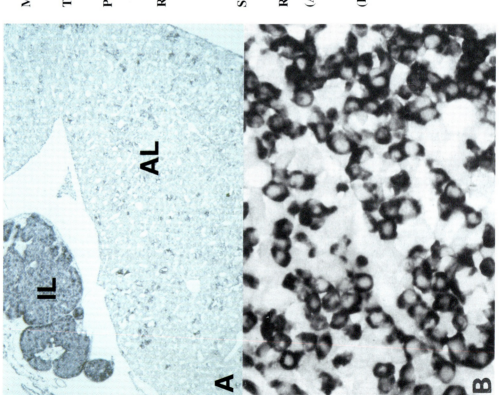

Method:
Immunohistology

Tissue:
Rat pituitary

Probe:
Oligonucleotide (30 mers)

Revelation:
Direct immunohistological reaction: anti-digoxigenin / alkaline phosphatase, revealed by NBT – BCIP

Signal:
Blue-violet precipitate

Preparation:
Paraffin-embedded sections

Labeling:
3′ extension by addition of dUTP–digoxigenin

Mode of observation:
Bright-field microscopy

Results:

(A): Revelation of mRNA coding for the adrenocorticotropic hormone (ACTH). The signal is present in all the cells of the intermediate lobe (IL). In the anterior lobe (AL), some cells are positive.

(B): Revelation of mRNA coding for the growth hormone (GH). The signal is present mainly in the cytoplasm of ≈50% of the cells in the anterior lobe.

Method:
Immunohistology

Tissue:
(A) Rat pituitary; (B) culture of Caski cells infected by the cytomegalovirus (CMV)

Probe:
(A) Oligonucleotide (30 mers);
(B) cDNA from CMV

Revelation:
Indirect immunohistological reaction:
(A): Anti-biotin; anti-species / colloidal gold, latensification with silver salts
(B): Anti-fluorescein; anti-species / peroxidase, revealed by AEC

Signal:
(A): Bright grains, counterstaining with toluidine blue
(B): Red precipitate, counterstaining with toluidine blue

Preparation:
(A) Frozen sections
(B) cytocentrifugation

Labeling:
(A) 3' extension by addition of dUTP – biotin; (B) nick translation incorporation of dUTP – fluorescein

Mode of observation:
(A) Epipolarization
(B) Bright-field microscopy

Results:
(A): Revelation of mRNA coding for the adrenocorticotropic hormone (ACTH) in all the cells in the intermediate lobe, and in some cells in the anterior lobe.
(B): Revelation of the CMV in the nuclei of all the cells. The CMV appears in the form of crystalline formations in the nuclei.

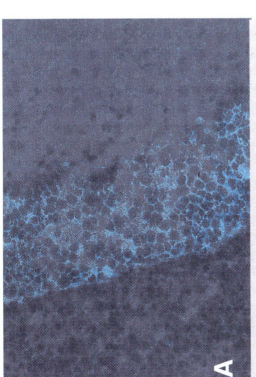
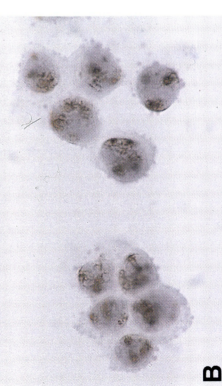

Examples of Observations

Method:
Immunohistology

Tissue:
Human condyloma

Probe:
cDNA in human papilloma virus
(**A**) HPV 6/11; (**B**) HPV 16/18

Revelation:
(**A**) Direct reaction: anti-fluorescein / alkaline phosphatase, revealed by NBT – BCIP; (**B**) Indirect reaction: anti-biotin, anti-species / alkaline phosphatase, revealed by NBT – BCIP

Signal:
Blue-violet precipitate

Preparation:
Paraffin-embedded sections

Labeling:
Nick translation with incorporation of: (**A**) dUTP – fluorescein; (**B**) dUTP – biotin

Mode of observation:
Bright-field microscopy

Results:

(**A**): Detection of HPV 6/11. Nuclei of superficial koilocitic cells are stongly positive, while nuclei of basal cells are negative (counterstained in red).

(**B**): Detection of HPV 16/18. Only the nuclei of some koilocytes are positive (without counterstaining).

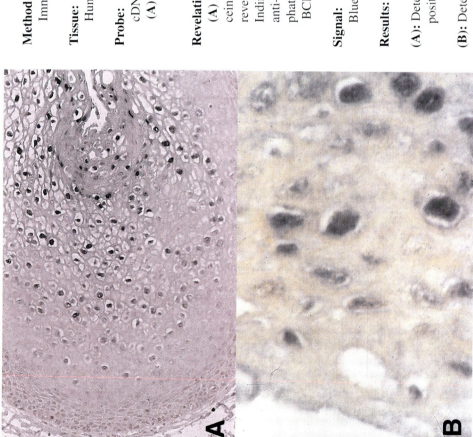

Examples of Observations

Method:
Immunohistology

Tissue: Hodgkin lymphoma

Probe:
Oligonucleotides (30 mers)

Revelation:
Indirect simultaneous reactions
• Anti-fluorescein, anti-species / alkaline phosphatase, revealed by NBT – BCIP

Signal:
Blue-violet precipitate (counterstaining: nuclear fast-red)

Preparation:
Paraffin-embedded sections

Labeling:
3' extension by addition of dUTP – fluorescein

Mode of observation:
Bright-field microscopy

Results:
Detection of mRNA of Epstein-Barr virus (EBER – Epstein-Barr early RNA):

(A): Positive in neoplastic cells in a Hodgkin lymphoma nodular esclersis type.

(B): A higher magnification shows in the center of the image a Reed-Sternberg cell with strong nuclear positivity.

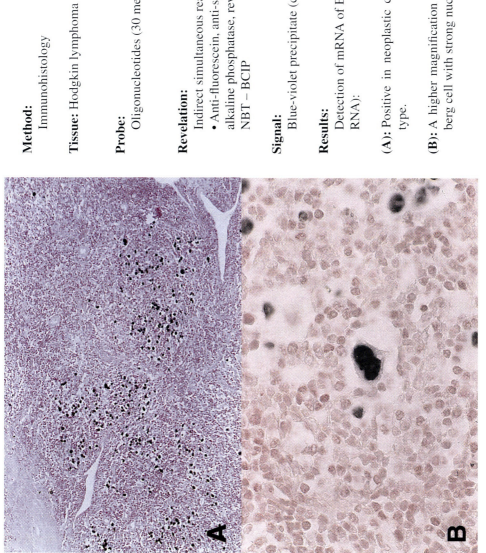

Examples of Observations

Method:
Immunohistology: double detection: **(A):** *in situ* hybridization + immunohistochemistry; **(B):** double *in situ* hybridization

Tissue:
(**A**) Undifferentiated nasopharyngeal carcinoma; (**B**) Rat pituitary

Probe:
Oligonucleotides (30 mers)

Preparation:
Paraffin-embedded sections

Labeling:
3' extension by addition of
(**A**) dUTP – fluorescein;
(**B**) dUTP – digoxigenin or dUTP – fluorescein

Revelation: Indirect simultaneous reactions:
(**A**) Immunohistochemistry: streptavidin – biotin / peroxidase complex, revealed by DAB (**B**) anti-digoxigenin, anti-species / alkaline phosphatase, revealed by NBT–BCIP and anti-fluorescein, anti-species/peroxidase, revealed by DAB

Mode of observation:
Bright-field microscopy

Signal:
(**A**): • Blue-violet precipitate for visualizing the hybrid labeled with fluorescein
• Brown precipitate for visualizing the cytokeratin immunoreactivity
(**B**): • Blue-violet precipitate for visualizing the hybrid labeled with digoxigenin
• Brown precipitate for visualizing the hybrid labeled with fluorescein

Results:
(**A**): Simultaneous localization of cytoplasmic cytokeratins (brown precipitate) and nuclear Epstein-Barr virus (blue and violet precipitate) in undifferentiated nasopharyngeal carcinoma.
(**B**): Simultaneous revelation, in the anterior lobe of the pituitary, of mRNA coding for the growth hormone (GH) (blue precipitate), and of mRNA coding for the adrenocorticotropic hormone (ACTH) (brown precipitate). The signals are present in different cells.

Appendices

Contents

A – EQUIPMENT

A 1 Practical Precautions	267
A 1.1 RNase-Free Conditions	267
A 1.2 Chemical Risks	268
A 1.3 Radioactive Risks	269
A 2 Sterilization	270
A 3 Pretreatment of Slides	270

B – REAGENTS

B 1 Water	273
B 1.1 Sterile	273
B 1.2 Diethylpyrocarbonate (DEPC)	273
B 2 Solutions	274
B 2.1 7.5 M Ammonium Acetate	274
B 2.2 3 M Sodium Acetate	274
B 2.3 DNA	275
B 2.4 RNA	275
B 2.5 Calcium Chloride	275
B 2.5.1 1 M Calcium Chloride	275
B 2.5.2 Calcium Chloride, Cobalt Chloride	276
B 2.6 4 M Lithium Chloride	276
B 2.7 1 M Magnesium Chloride	276
B 2.8 5 M Sodium Chloride	277
B 2.9 Denhardt's Solution 50X	277
B 2.10 1 M Dithiotreitol (DTT)	278
B 2.11 DNase I	278
B 2.12 500 mM Ethylene Diamine Tetra-acetic Acid (EDTA)	278
B 2.13 Deionized Formamide	279
B 2.14 Poly (A)	279
B 2.15 Polyethylene Glycol	279
B 2.16 Pronase	280
B 2.17 Proteinase K	280
B 2.18 RNase A	281
B 2.19 Saccharose	281
B 2.20 Sarcosyl	281
B 2.21 Caustic Soda	282

B 2.22 Dextran Sulfate	282
B 2.23 1 *M* Tris	282
B 2.24 Triton X-100	283
B 3 Buffers	283
B 3.1 100 m*M* Acetylation Buffer	283
B 3.2 DNase I Buffer 10X	283
B 3.3 Phosphate Buffer	284
B 3.3.1 1 *M* Phosphate	284
B 3.3.2 PBS	284
B 3.4 SSC Buffer 20X	285
B 3.5 TE (Tris – EDTA) Buffer	286
286	286
B 3.5.2 TE/NaCl Buffer 10X	287
B 3.6 Tris - HCl Buffer	287
B 3.6.1 100 m*M* Tris – HCl Buffer	287
B 3.6.2 Tris – HCl/CaCl2 Buffer	288
B 3.6.3 Tris – HCl/EDTA Buffer	288
B 3.6.4 Tris – HCl/Glycine Buffer	288
B 3.6.5 Tris – HCl/NaCl Buffer	289
B 3.6.6 Tris – HCl/NaCl / MgCl2 Buffer	289
B 4 Fixatives	289
B 4.1 Acetone	290
B 4.2 Alcohol	290
B 4.2.1 Ethyl Alcohol	290
B 4.2.2 Methyl Alcohol	290
B 4.3 Formol	291
B 4.4 Paraformaldehyde	291
B 4.4.1 Paraformaldehyde (40%)	291
B 4.4.2 Paraformaldehyde (4%)	292
B 4.4.3 Paraformaldehyde 4%/Glutaraldehyde 0.05%	293
B 5 Autoradiography	293
B 5.1 Standard Developer	293
B 5.2 Fixative	294
B 6 Immunocytology	294
B 6.1 Blocking Solutions	294
B 6.1.1 Nonspecific Sites	294
B 6.1.2 Endogenous Alkaline Phosphatases	294
B 6.1.3 Endogenous Peroxidases	295
B 6.2 Chromogens	295
B 6.2.1 Alkaline Phosphatase	295
B 6.2.2 Peroxidase	296
B 7 Stains	298
B 7.1 Toluidine Blue	298
B 7.2 Cresyl Violet	298
B 7.3 Eosin	299
B 7.4 Harris's Hematoxylin	299
B 7.5 Rapid Nuclear Red	300
B 7.6 Methyl Green	300

B 8	Mounting Media...	300
	B 8.1 Aqueous...	300
	B 8.1.1 Buffered Glycerine.............................	301
	B 8.1.2 Moviol.......................................	301
	B 8.2 Permanent..	301
	B 8.3 For Fluorescence.....................................	302

Appendix A
EQUIPMENT

This appendix presents the procedures to be followed, and the precautions to be taken, in preparing a particular substance for *in situ* hybridization in the best possible conditions of preservation, not only of morphology but also of the nucleic acids to be revealed.

A1 PRACTICAL PRECAUTIONS

A1.1 RNase-Free Conditions

❑ *Working Areas*
- Well marked-out benchtop area

  ↝ Receptacles used for radioactivity can also be used to localize manipulation sites with high precision.

- Cleanliness of sites close to sterility

  ↝ There are reagents that destroy RNases.

❑ *Probe*
- Use of cRNA probes

  ↝ RNase activity will be inhibited in the buffers used (presence of RNasin).
  ↝ Commercial solutions for inhibiting RNases are becoming available.

- Storage
  — DEPC-treated water
  — Choice of TE buffer (TE / NaCl)

  ↝ This buffer, which is very stable, inhibits the action of enzymes that break down DNA.

Equipment

❏ *Tissue*
- Sampling in sterile conditions
- Storage of slides — Place the dry slides in hermetic boxes containing a desiccant (e.g., silica gel)

↪ Drying is indispensable before the storage of the slides at –20°C: RNases and DNases function only in the presence of water, so that storage in the absence of water is the best inhibitor of RNases. This storage can preserve the sections in a state of hybridization for up to several years.

- Opening of the storage boxes in suitable conditions of humidity

↪ Wait at least 2 h at room temperature before opening the boxes (to reduce risk of condensation and rehydration of the sections).

❏ *Equipment/Reagents/Solutions*
- Eppendorf tubes
- Sterilized cones
- Gloves

↪ Disposable equipment is preferable to sterilized equipment.
↪ Do not directly touch the equipment or the reagents.

- New equipment

↪ Use receptacles reserved exclusively for these manipulations.

- Receptacles

↪ Use receptacles reserved for these manipulations, and sterilize them immediately after use (otherwise, use disposable equipment).

- Water treated with DEPC or containing RNase inhibitors

↪ The risk of contamination increases with time, in relation to frequency of opening.
↪

❏ *Conclusion*
Preventing the appearance of RNases in the first place is easier than getting rid of them afterwards.

A1.2 Chemical Risks

❏ *Reagents/Solutions*
- Formaldehyde
- Formamide
- Glutaraldehyde
- Organic solvents (xylene, acetone, etc.)

↪ Follow the manufacturers' instructions.

❏ *Practical Precautions*
All these reagents are dangerous.

↪ Use in ventilated spaces.

❏ *Storage/Waste Products*
Special containers of variable size are used to collect glassware, chemical residues, and incineratable biological materials.

A1.3 Radioactive Risks

Radioactive hazards can have a number of different origins:

- Gloves

- Protection / control

- Storage of radioactive sources
- Control and elimination of waste products

- Radioprotection training courses

⇝ Contact the person responsible.

⇝ Change regularly to prevent contamination. Gloves are not, in themselves, a barrier to irradiation.
⇝ Regularly check surfaces, screens, hands, and equipment (^{35}S, ^{33}P, ^{32}P).
⇝ Store in receptacles kept in a special room.
⇝ These are removed in accordance with the norms imposed by the hygiene and safety legislation in force.
⇝ Familiarity with safety conditions is a necessity.

Prevention During the Utilization of Radioelements			
Isotopes	Wearing of Dosifilm	Special Equipment[a]	Risks, Controls (appropriate detector)
^{3}H	No	• Screen: none • Two pairs of gloves to be worn	• Contamination • No irradiation
^{35}S ^{33}P	No	• Screens: either 0.2 mm glass or 0.3 mm Plexiglas • Two pairs of gloves to be worn	• No irradiation • Contamination
^{32}P	Yes	• Screens: either 4 mm glass or 8 mm Plexiglas; Never a lead screen (generator of secondary radiation) • Two pairs of gloves to be worn (with frequent changes)	• Irradiation • Contamination
Decontamination, Control, Elimination of Waste:			
Leave the equipment in a diluted solution of decontaminant, then rinse abundantly in water. After decrease (>10 half-lives) and controls, the waste material can be disposed of as ordinary waste.			

[a] Specially adapted working areas with signs, extractor fans, protective equipment (screen, gloves, benchtop, Plexiglas bin).

A2 STERILIZATION

The sterilization of solutions and of small equipment is indispensable in *in situ* hybridization.

❏ *Equipment*
- Aluminum foil
- Autoclave
- Indicator of sterilization
- Oven

❏ *Minor Equipment*
- Glass staining trays

❏ *Protocol*
Treatment of equipment, including magnetic bars, in:
- Poupinel oven   **2 h at 180°C**
- Autoclave   **2 h at 105°C** or **30 min at 125°C**

↪ Maintain conditions of cleanliness, and use solutions and equipment exclusively for *in situ* hybridization.

↪ The color of an indicator changes after sterilization.

↪ Glass equipment is autoclaved or sterilized in a Poupinel oven.

↪ It has been shown that RNases can partly resist such treatment.

↪ Two bars. Allow the equipment to cool in the oven (there is a risk of breakage if it is placed on a cold surface).

A3 PRETREATMENT OF SLIDES

❏ *Equipment*
- Metal pincers
- Oven or autoclave
- Trays, slide-holders

❏ *Reagents*
- Acetone
- Alcohol 95%
- 3-Amino-propyl-tri-ethoxy-silane
- Hydrochloric acid 10 N
- Sterile distilled water

❏ *Solutions*
- Cleaning: Alcohol / HCl (5 mL HCl for 1 L alcohol 95%)

↪ *See* Appendix B1.1.

↪ This is a stable solution which can be made in bulk and saved.

- Treatment: 3-Amino-propyl-tri-ethoxy-silane at 2% in acetone

⇒ The solution is unstable; it must be prepared at time of use.

❑ *Precaution*

The slides should be manipulated with gloves from the beginning to the end of the treatment.

❑ *Protocol*
1. Wash
    - Alcohol / HCl   **Overnight**
    - Running water   **1 h**
    - Distilled water   **1 min**
2. Dry the slides in the oven   **180 or 42°C for 15 to 60 min**

⇒ Sterilization at 180°C is optional.

3. Allow to cool
4. Immerse in the treatment solution **5 to 15 s**
5. Wash
    - Acetone   **2 × 1 min**
    - Sterile water   **1 min**
6. Dry   **Overnight at 42°C**

❑ *Storage*

Storage is possible for up to a year at room temperature and in dust-free conditions.

Appendix B
Reagents

B1 WATER

B1.1 Sterile

❏ *Reagents*
• Distilled water
❏ *Precaution*
Do not save after opening.
❏ *Protocol*
Sterilize in an autoclave. **2 h at 105°C**

❏ *Storage*
Store at room temperature.

⇝ Sterile water is sufficient for the detection of double-stranded DNA (e.g., viruses).

B1.2 Diethylpyrocarbonate (DEPC)

❏ *Reagents/Solutions*
• DEPC
• Distilled water
❏ *Precaution*
DEPC is a dangerous reagent.
❏ *Protocol*
1. Mix
    • DEPC **0.5 to 1 mL**
    • Water **1000 mL**
2. Shake under a ventilation hood. **Overnight**
3. Sterilize in an autoclave. **30 min at 105°C**

❏ *Storage*
Store at room temperature.

⇝ DEPC is **indispensable** with cRNA probes.

B2 SOLUTIONS

B2.1   7.5 M Ammonium Acetate   ↝ Stock solution

❏ *Reagents/Solutions*
- Ammonium acetate
- Hydrochloric acid 10 N
- Sterile water   ↝ *See* Appendix B1.1.

❏ *Precautions*
Check the nonhydrated state of the powder.
This is a volatile reagent.

❏ *Protocol*
- Ammonium acetate   57.81 g   ↝ MW = 77.08.
- Sterile water   80 mL

1. Mix.
2. Adjust the pH to 5.5 with HCl:
   - Sterile water.   to 100 mL

❏ *Sterilization*
Sterilize in an autoclave.   2 h
                                at 105°C

❏ *Storage*
- At –20°C in aliquots of 50 to 100 µL
- At room temperature for some months

B2.2   3 M Sodium Acetate   ↝ Stock solution

❏ *Reagents/Solutions*   ↝ AR quality, to be used only for *in situ* hybridization

- Glacial acetic acid
- Sodium acetate   ↝ CH_3COONa
- Sterile water   ↝ *See* Appendix B1.1.

❏ *Precaution*
None.

❏ *Protocol*
- Sodium acetate   24.6 g   ↝ MW = 82.03.
- Sterile water   80 mL

1. Mix with magnetic stirring.
2. Adjust the pH to 4.8 with acetic acid:
   - Sterile water   to 100 mL

❏ *Sterilization*
Sterilize in an autoclave.   2 h
                                at 105°C

❏ *Storage*
Store at room temperature for some months, in aliquots of 50 to 100 µL.

B2.3 DNA

❏ *Reagents/Solutions*
- Herring sperm DNA
- Salmon sperm DNA
- Sterile water

❏ *Precaution*
There is risk of bacterial contamination.

❏ *Protocol*
- DNA   **10 mg**
- Sterile water   **to 1 mL**

Sonicate for 10 min.

❏ *Storage*
Store at –20°C.

↪ **Stock solution** 10 mg / mL sterile water

↪ *See* Appendix B1.1.

↪ *See* Appendix B1.1.
↪ Use maximum power.

B2.4 RNA

❏ *Reagents/Solutions*
- DEPC-treated water
- Yeast tRNA

❏ *Precaution*
There is risk of hydrolysis.

❏ *Protocol*
- RNA   **10 mg**
- DEPC-treated water   **to 1 mL**

Sonicate 10 min.

❏ *Storage*
Store at –20°C in aliquots of 50 µL.

↪ **Stock solution** 10 mg/mL sterile water

↪ *See* Appendix B1.2.

↪ Use maximum power.

B2.5 Calcium Chloride

B2.5.1 1 M calcium chloride

❏ *Reagents/Solutions*

- Calcium chloride
- Sterile water

❏ *Precaution*
None.

❏ *Protocol*
- Calcium chloride   **14.7 g**
- Sterile water   **to 100 mL**

❏ *Sterilization*
Sterilize in an autoclave.   **2 h at 105°C**

❏ *Storage*
Store at room temperature.

↪ **Stock solution**
↪ AR quality, to be used only for *in situ* hybridization
↪ $CaCl_2 \cdot 2H_2O$
↪ *See* Appendix B1.1.

↪ MW = 147.02.

B2.5.2 Calcium chloride/cobalt chloride

❏ *Reagents/Solutions*
- Calcium chloride
- Cobalt chloride
- Sterile water

❏ *Precaution*
None.

❏ *Protocol*
- Calcium chloride   0.029 g
- Cobalt chloride   0.025 g
- Sterile water   to 100 mL

❏ *Sterilization*
Sterilize in an autoclave.   2 h
                         at 105°C

❏ *Storage*
Store at room temperature.

↪ **Stock solution** 2 mM $CaCl_2$ / 2 mM $CoCl_2$

↪ $CaCl_2 \cdot 2H_2O$
↪ $CoCl_2$
↪ *See* Appendix B1.1.

↪ MW = 147.02.
↪ MW = 129.83.

B2.6   4 M Lithium Chloride

❏ *Reagents/Solutions*

- Lithium chloride
- Sterile water

❏ *Precaution*
There is risk of bacterial contamination.

❏ *Protocol*
- Lithium chloride   8.48 g
- Sterile water   to 50 mL
1. Mix.
2. Filter on a 0.22-μm filter.

❏ *Storage*
Store at 4°C for some weeks.

↪ **Stock solution**

↪ AR quality, to be used only for *in situ* hybridization.
↪ LiCl
↪ *See* Appendix B1.1.

↪ MW = 42.39.

B2.7   1 M Magnesium Chloride

❏ *Reagents/Solutions*

- Magnesium chloride
- Sterile water

❏ *Precaution*
None.

❏ *Protocol*
- Magnesium chloride   20.3 g
- Sterile water   to 100 mL

❏ *Sterilization*
Sterilize in an autoclave.   2 h
                         at 105°C

↪ **Stock solution**

↪ AR quality, to be used only for *in situ* hybridization.
↪ $MgCl_2 \cdot 6\ H_2O$
↪ *See* Appendix B1.1.

↪ MW = 203.30.

Appendix B

❏ *Storage*
Store at room temperature.
❏ *Precaution*
None.

B2.8 5 M Sodium Chloride

❏ *Reagents/Solutions*

- Sodium chloride
- Sterile water

❏ *Precaution*
None.
❏ *Protocol*
- Sodium chloride   14.6 g
- Sterile water   to 50 mL

❏ *Sterilization*
Sterilize in an autoclave.   2 h
                              at 105°C

❏ *Storage*
Store at room temperature.

↪ Stock solution

↪ AR or molecular biology quality, to be used only for *in situ* hybridization
↪ NaCl
↪ *See* Appendix B1.1.

↪ MW = 58.44.

B2.9 Denhardt's Solution 50X

❏ *Reagents/Solutions*
- Bovine serum albumin (fraction V)
- Ficoll 400
- Polyvinylpyrzolidone
- Sterile water

❏ *Precaution*
There is risk of bacterial contamination.
❏ *Protocol*
1. Add:
   - BSA   1 g
   - Ficoll 400   1 g
   - Polyvinylpyrzolidone   1 g
   - Sterile water   to 100 mL
2. Leave the mixture to hydrate overnight before shaking.
3. Shake the mixture gently and intermittently for some days.

❏ *Storage*
Store at –20°C in aliquots.

↪ Stock solution

↪ BSA, 5 times crystallized

↪ PVP
↪ *See* Appendix B1.1.

↪ Solution can be frozen and thawed.

B2.10   1 M Dithiotreitol (DTT)

❏ *Reagents/Solutions*
- Dithiotreitol
- Sterile water

❏ *Precaution*
DTT is a very volatile reagent, unstable at room temperature. It does not resist denaturation.

❏ *Protocol*
1. In a sterile Eppendorf tube, mix:
    - Dithiotreitol   1.54 g
    - Sterile water   10 mL

❏ *Storage*
Store at –20°C in aliquots of 100 µL.

↪ Stock solution
↪ Molecular biology quality
↪ $C_4H_{10}O_2S_2$
↪ See Appendix B1.1.

↪ MW = 154.24.

↪ DTT must never be refrozen. Its foul smell is the best guarantee of its condition.

B2.11   DNase I

❏ *Reagents/Solutions*
- DNase I
- Sterile water

❏ *Precaution*
There is risk of RNase contamination.

❏ *Protocol*
Dissolve:
- DNase I   1 mg
- Sterile water   1 mL

❏ *Storage*
Store at –20°C in aliquots of 100 µL.

↪ Stock solution 1 mg / mL
↪ Molecular biology quality.
↪ Check the activity of the enzyme.
↪ See Appendix B1.1.

↪ Express in units.

B2.12   500 mM Ethylene Diamine Tetra-acetic Acid (EDTA)

❏ *Reagents/Solutions*

- Caustic soda 10 N
- EDTA
- Sterile water

❏ *Precaution*
Reagent is toxic.

❏ *Protocol*
1. Dissolve:
    - EDTA   186 g
    - Sterile water   to 1 L
2. Adjust the pH to 8.0 with caustic soda.

↪ Stock solution

↪ AR quality, to be used only for *in situ* hybridization
↪ See Appendix B2.21.
↪ $C_{10}H_{14}N_2O_8Na_2 \cdot 2H_2O$ or Titriplex III
↪ See Appendix B1.1.

↪ MW = 372.24.

❏ *Sterilization*
Sterilize in an autoclave. **2 h at 105°C**

❏ *Storage*
Store at room temperature, in aliquots.

B2.13 Deionized Formamide

❏ *Equipment*
- Sterile flask
- Whatman filter No. 1M

❏ *Reagents/Solutions*   ↪ Molecular biology quality
- Amberlite resin, 20 to 50 mesh
- Formamide   ↪ CH_3NO

❏ *Precaution*
Avoid contact.

❏ *Protocol*   ↪ Deionized formamide exists in a commercial solution.

- Amberlite **5 g**
- Formamide filter **50 mL**   ↪ MW = 45.04.
1. Shake gently. **30 min**
2. Filter on Whatman paper.

❏ *Storage*
Store at –20°C in sterile 1-mL tubes, in light-free conditions.
↪ Solution must never be stored unfrozen.
↪ If it is liquid at –20°C, it must not be used.

B2.14 Poly (A)

↪ **Stock solution** 10 mg / mL

❏ *Reagents/Solutions*
- DEPC-treated water   ↪ *See* Appendix B1.2.
- Poly A   ↪ $C_{11}H_{17}N_4O_{12}$ (polyribonucleotide adenylate)

❏ *Precaution*
There is risk of hydrolysis by RNase.

❏ *Storage*
Store at –20°C.

B2.15 Polyethylene Glycol

↪ **Stock solution** 10 mg / mL

❏ *Reagents/Solutions*
↪ AR quality, to be used only for *in situ* hybridization.
- Polyethylene glycol (PEG)   ↪ $HO[C_2H_4O]_nH$
- Sterile water   ↪ *See* Appendix B1.1.

❏ *Precaution*
Avoid contact.

❏ *Protocol*
1. In a sterile Eppendorf tube
- Polyethylene glycol   **10 mg**   ↪ MW = 5000 to 7000.
- Sterile water   **1 mL**
2. Mix.
❏ *Storage*
Store at –20°C in aliquots of 50 to 100 µL.

B2.16 Pronase

↪ **Stock solution** 10 mg / mL

❏ *Reagents/Solutions*
- Pronase
- Sterile water   ↪ *See* Appendix B1.1.
- TE buffer   ↪ *See* Appendix B3.5.1.

❏ *Precaution*
Proteolysis occurs at room temperature.
❏ *Protocol (stock solution)*
- Pronase   **10 mg**
- Sterile water   **1 mL**

❏ *Storage*
Store at –20°C in aliquots of 250 µL.
❏ *Protocol (working solution)*
1. Dilute in TE buffer at time of use.
2. Predigest.   **2 h at 37°C**

B2.17 Proteinase K

↪ **Stock solution** 10 mg / mL

❏ *Reagents/Solutions*
- Proteinase K
- Sterile water   ↪ *See* Appendix B1.1.

❏ *Precaution*
Weak proteolysis occurs at room temperature.
❏ *Protocol*
1. In a sterile Eppendorf tube
   - Proteinase K   **10 mg**
   - Sterile water   **1 mL**
2. Mix.
❏ *Storage*
Store at –20°C in aliquots of 100 µL.

↪ Dilute in Tris–HCl/$CaCl_2$ buffer.
↪ *See* Appendix B3.6.2.

Appendix B

B2.18 RNase A

❏ *Reagents/Solutions*
• RNase A
• Sterile water
• TE / NaCl buffer

❏ *Precaution*
There is risk of DNase contamination.

❏ *Protocol*
• RNase A 100 μg
• Sterile water 1 mL

❏ *Storage*
Store at –20°C in aliquots of 40 μL.
Store at 0°C, lyophilized reagent.

↪ **Stock solution** 100 μg / mL

↪ Molecular biology quality
↪ Bovine pancreas ribonuclease
↪ *See* Appendix B1.1.
↪ *See* Appendix B3.5.2.

B2.19 Saccharose

❏ *Reagents/Solutions*
• Phosphate or PBS buffer
• Saccharose

❏ *Precaution*
There is risk of contamination.

❏ *Protocol*
• Saccharose 30 or 50 g
• Buffer to 100 mL

❏ *Storage*
It is preferable to store the solution at 4°C.

↪ **Stock solution,** 30 or 50%.

↪ *See* Appendix B3.3.
↪ $C_{12}H_{22}O_{11}$

↪ MW = 342.34.

B2.20 Sarcosyl

❏ *Reagents*
• Sarcosyl
• Sterile water

❏ *Precaution*
Sacrosyl is toxic.

❏ *Protocol*
• Sarcosyl 20 mL
• Sterile water 80 mL

❏ *Storage*
Store at 4°C for some weeks.

↪ **Stock solution** 20%

↪ *See* Appendix B1.1.

B2.21 Caustic Soda

❏ *Reagents/Solutions*
- Caustic soda
- Sterile water

❏ *Precaution*
Caustic soda is an extremely corrosive reagent.

❏ *Protocol*
- Caustic soda   **40 g**
- Sterile water   **to 100 mL**

Mix slowly.

❏ *Storage*
Store at room temperature.

↪ **Stock solution** 10 N

↪ AR quality
↪ NaOH
↪ *See* Appendix B1.1.

↪ MW = 40.00.

B2.22 Dextran Sulfate

❏ *Reagents/Solutions*
- Dextran sulfate
- Sterile water

❏ *Precaution*
This reagent is very difficult to pipette. Prepare only the quantity necessary.

❏ *Protocol*
1. In a sterile Eppendorf tube:
   - Dextran sulfate   **50 mg**
   - Sterile water   **65 µL**
2. Mix.
3. Microcentrifuge for a few seconds.

❏ *Storage*
Store at –20°C, in aliquots of 100 µL, for some months.

↪ **Stock solution** 50%

↪ Molecular biology quality

↪ *See* Appendix B1.1.

↪ The other components of the hybridization medium will be added to the aliquots.

B2.23 1 M Tris

❏ *Reagents/Solutions*
- Hydrochloric acid 10 N
- Sterile water
- Tris – hydroxymethyl – aminomethane

❏ *Precaution*
None.

❏ *Protocol*
Dissolve:
- Tris base   **121 g**
- Sterile water   **to 1 L**

❏ *Storage*
Store at room temperature, for some weeks.

↪ **Stock solution**

↪ *See* Appendix B1.1.
↪ $C_4H_{11}NO_3$, Tris base in powder

↪ MW = 121.16.

B2.24 Triton X-100

↬ **Stock solution** 0.1 to 0.3%

❑ *Reagents/Solutions*
- Sterile water
- Triton X-100

❑ *Precaution*
Reagent is difficult to pipette.

❑ *Protocol*
Dissolve:
- Triton X-100   0.1 to 0.3 g

- Sterile water   to 100 mL

❑ *Storage*
Store at room temperature.

↬ AR quality
↬ *See* Appendix B1.1.
↬ $C_{34}H_{62}O_{11}$

↬ MW = 646.87.
↬ Triton is a detergent that is difficult to pipette with exactitude.

B3 BUFFERS

B3.1 100 mM Acetylation Buffer

↬ 100 mM triethanolamine buffer; pH 8.0

❑ *Reagents/Solutions*
- Hydrochloric acid 10 N
- Sterile water
- Triethanolamine

❑ *Precaution*
Buffer is toxic.

❑ *Protocol*
- Triethanolamine   3.32 mL
- Sterile water   200 mL
Adjust to pH 8.0
- HCl   ≈ 1 mL
- Sterile water   to 250 mL

❑ *Storage*
Store at 4°C.

↬ AR quality

↬ *See* Appendix B1.1.
↬ $C_6H_{15}NO_3$ (Tris (hydroxy-2-ethyl)-amine)

↬ MW = 149.19.

B3.2 DNase I Buffer 10X

↬ Stock solution

❑ *Reagents/Solutions*
- Bovine serum albumin
- Hydrochloric acid 10 N
- 500 mM magnesium chloride

↬ AR quality
↬ BSA 5X, crystallized

↬ *See* Appendix B2.7.

Reagents

- Sterile water
- 1 M Tris

❏ *Precaution*

There is risk of RNase contamination; must be prepared in sterile conditions.

❏ *Protocol*

- Tris – HCl, pH 7.6   **500 mM**
- MgCl$_2$ **100 mM**
- BSA   **0.5 mg /mL**

❏ *Storage*

Store at –20°C in aliquots.

↪ *See* Appendix B1.1.
↪ *See* Appendix B2.23.

↪ *See* Appendix B3.6.1.

B3.3 Phosphate Buffer

B3.3.1 1 M phosphate

❏ *Reagents/Solutions*
- Bisodium phosphate

- Monosodium phosphate
- Sterile water

❏ *Precaution*

Preparations obtained in sachets should be transferred to sterile receptacles, with sterile water.

❏ *Protocol*

1. Dissolve:
   - Bisodium phosphate   **14.19 g**
   - Monosodium phosphate   **13.8 g**
   - Sterile water   **to 100 mL**
2. Check.   **pH 7.4**

❏ *Sterilization*

Sterilize in an autoclave.   **2 h at 105°C**

❏ *Storage*

Store at room temperature, or 4°C.

↪ Stock solution

↪ AR quality
↪ Na$_2$HPO$_4$. Hydrated reagents can be used, taking into account variations in molar mass.
↪ NaH$_2$PO$_4$·H$_2$O
↪ *See* Appendix B1.1.

↪ For 100 mL

↪ MW = 141.96.
↪ MW = 138.00.

B3.3.2 PBS

❏ *Reagents/Solutions*

- Bisodium phosphate
- Caustic soda 10 N
- Monopotassium phosphate
- Potassium chloride
- Sodium chloride

↪ Phosphate buffer saline
↪ 10 mM mono-di-phosphate buffer; 150 mM NaCl; 10 mM KCl; pH 7.4
↪ Can be prepared in stock solution 10X
↪ AR quality, to be used only for *in situ* hybridization
↪ Na$_2$HPO$_4$·12H$_2$O
↪ *See* Appendix B2.21.
↪ KH$_2$PO$_4$·3H$_2$O
↪ KCl
↪ NaCl
↪ *See* Appendix B2.8.

- Sterile water ↪ See Appendix B1.1.
❏ *Precaution*
There are risks of contamination.
❏ *Protocol*
1. Add:
   - Sodium chloride   8.76 g   ↪ MW = 58.44.
   - Potassium chloride   0.74 g   ↪ MW = 74.56.
   - Bisodium phosphate   3.58 g   ↪ MW = 358.14.
   - Monopotassium phosphate   1.90 g   ↪ MW = 190.15.
   - Sterile water   to 1 L
2. Adjust the pH with caustic soda.   7.4

❏ *Sterilization*
Sterilize in an autoclave.   2 h
                                      at 105°C
❏ *Storage*
Store at room temperature before opening, then at 4°C.

B3.4 SSC Buffer 20X

↪ Standard saline citrate
↪ **Stock solution**
↪ 3 M sodium chloride; 300 mM sodium citrate; pH 7.0

❏ *Reagents/Solutions*
- Caustic soda 10 N ↪ AR quality
- Sodium chloride   ↪ See Appendix B2.21.
- Sodium citrate   ↪ NaCl
- Sterile water   ↪ $C_6H_5Na_3O_7 \cdot 2H_2O$
❏ *Precaution*   ↪ See Appendix B1.1.
None.

❏ *Protocol*
1. Add:
   - Sodium chloride   175.3 g   ↪ MW = 58.44.
   - Sodium citrate   88.2 g   ↪ MW = 294.10.
   - Sterile water   to 1 L
2. Adjust the pH with caustic soda.   7.0

❏ *Sterilization*
Sterilize in an autoclave.   2 h
                                      at 105°C
❏ *Storage*
Store at room temperature before opening, then at 4°C.

B3.5 TE (Tris–EDTA) buffer

❏ Reagents

- EDTA

- Tris (hydroxymethyl – aminomethane)

↪ Tris/EDTA

↪ AR quality, to be used only for *in situ* hybridization
↪ $C_{10}H_{14}N_2O_8Na_2 \cdot 2H_2O$ (ethylene diamine tetra-acetic acid, or Titriplex III)
↪ $C_4H_{11}NO_3$

B3.5.1 TE buffer 10X

❏ Reagents/Solutions
- EDTA

- Hydrochloric acid 10 N
- Sterile water
- Tris

❏ Precaution
EDTA is toxic.

❏ Protocol
1. Add:
   - Tris   12.1 g
   - EDTA   3.7 g
   - Sterile water   to 1 L
2. Adjust the pH with HCl.   7.6

❏ Sterilization
Sterilize in an autoclave.   2 h at 105°C

❏ Storage
Store at room temperature before opening, then at 4°C.

↪ Stock solution
↪ 100 mM Tris–HCl; 10 mM EDTA; pH 7.6

↪ Powder, or 500 mM stock solution (*see* Appendix B2.12)

↪ *See* Appendix B1.1.
↪ $C_4H_{11}NO_3$ or 1 M stock solution (*see* Appendix B2.23)

↪ MW = 121.16.
↪ MW = 372.24.

B3.5.2 TE/NaCl buffer 10X

❏ Reagents/Solutions
- Hydrochloric acid 10 N
- Sodium chloride
- Sterile water
- TE buffer 10X

❏ Precaution
EDTA is toxic.

↪ Stock solution
↪ 100 mM Tris – HCl; 10 mM EDTA; 5 M NaCl; pH 7.6
↪ AR quality

↪ NaCl
↪ *See* Appendix B1.1.
↪ *See* Appendix B3.5.
↪

□ *Protocol*
1. Add:
   • Sodium chloride   292.2 g   ↪ MW = 58.44.
   • TE buffer 10X   to 1 L
2. Adjust the pH with HCl.   7.6
□ *Sterilization*
Sterilize in an autoclave.   2 h
                             at 105°C

□ *Storage*
Store at room temperature before opening, then at 4°C.

B3.6 Tris – HCl Buffer

B3.6.1 100 mM Tris – HCl Buffer

□ *Reagents/Solutions*   ↪ AR quality
• Hydrochloric acid 10 N
• Sterile water   ↪ See Appendix B1.1.
• 1 M Tris   ↪ See Appendix B2.23.
□ *Precaution*
There is risk of contamination by exogenous organisms.
□ *Protocol*
1. Add:
   • Tris   1 vol
   • Sterile water   9 vol
2. Adjust the pH with HCl.   7.6
□ *Sterilization*   ↪ Impossible, or very difficult.
□ *Storage*
Store at 4°C.

B3.6.2 Tris – HCl/CaCl$_2$ buffer   ↪ 20 mM Tris–HCl; 2 mM CaCl$_2$; pH 7.6

□ *Reagents/Solutions*   ↪ AR quality
• 1 M calcium chloride   ↪ See Appendix B2.5.1.
• Hydrochloric acid 10 N
• Sterile water   ↪ See Appendix B1.1.
• 1 M Tris   ↪ See Appendix B2.23.
                                        ↪

□ *Precaution*
There is risk of contamination.

□ *Protocol*
1. Add:
   • Tris   20 mL
   • Calcium chloride   2 mL

2. Adjust the pH with HCl.   **7.6**
- Sterile water   **to 1 L**

❏ *Sterilization*   ↪ Impossible, or at least very difficult.
❏ *Storage*
Store at 4°C.

B3.6.3 Tris – HCl/EDTA buffer

↪ 10 mM Tris – HCl; 1 mM EDTA; pH 8.0

❏ *Reagents/Solutions*   ↪ AR quality
- 500 mM EDTA   ↪ *See* Appendix B2.12.
- Hydrochloric acid 10 N
- Sterile water   ↪ *See* Appendix B1.1.
- 1 M Tris   ↪ *See* Appendix B2.23.

❏ *Precaution*
There is risk of contamination by exogenous organisms.

❏ *Protocol*
1. Add:
    - 1 M Tris   **10 mL**
    - 500 mM EDTA   **2 mL**
    - Sterile water   **to 1 L**
2. Adjust the pH with HCl.   **8.0**

❏ *Sterilization*   ↪ Impossible, or at least very difficult.
❏ *Storage*
Prepare at time of use.

B3.6.4 Tris – HCl/glycine buffer

↪ 50 mM Tris – HCl; 50 mM glycine; pH 7.4

❏ *Reagents/Solutions*   ↪ AR quality
- Glycine   ↪ $C_2H_5NO_2$
- Hydrochloric acid 10 N
- Sterile water   ↪ *See* Appendix B1.1.
- 1 M Tris   ↪ *See* Appendix B2.23.

❏ *Precaution*
There is risk of contamination by exogenous organisms.

❏ *Protocol*
1. Add:
    - Glycine   **0.37 g**   ↪ MW = 75.07.
    - Tris   **5 mL**
    - Sterile water   **to 100 mL**
2. Adjust the pH with HCl.   **7.4**

❏ *Storage*
- At –20°C in aliquots
- At 4°C

B3.6.5 Tris – HCl/NaCl buffer

↪ 50 mM Tris – HCl; 600 mM NaCl; pH 7.6

❑ *Reagents/Solutions*
- Hydrochloric acid 10 N ↝ AR quality
- 5 M sodium chloride   ↝ *See* Appendix B2.8.
- Sterile water   ↝ *See* Appendix B1.1.
- 1 M Tris   ↝ *See* Appendix B2.23.

❑ *Precaution*
There is risk of contamination by exogenous organisms.

❑ *Protocol*
1. Add:
   - 5 M sodium chloride   12 mL
   - 1 M Tris   5 mL
   - Sterile water   to 100 mL
2. Adjust the pH with HCl.   7.6

❑ *Storage*
Prepare at time of use.

B3.6.6 Tris – HCl/NaCl/MgCl$_2$ buffer

↝ 50 mM Tris – HCl; 300 mM NaCl; 50 mM MgCl$_2$; pH 9.5

❑ *Reagents/Solutions*
- Hydrochloric acid 10 N ↝ AR quality
- 1 M magnesium chloride   ↝ *See* Appendix B2.7.
- 5 M sodium chloride   ↝ *See* Appendix B2.8.
- Sterile water   ↝ *See* Appendix B1.1.
- 1 M Tris   ↝ *See* Appendix B2.23.

❑ *Precaution*
There is risk of contamination by exogenous organisms.

❑ *Protocol*
1. Add:
   - 1 M Tris   50 mL
   - 5 M NaCl   60 mL
   - 1 M MgCl$_2$ 50 mL
2. Adjust to pH 9.5 with HCl.
   - Sterile water   to 1 L

❑ *Storage*
Prepare at time of use.

B4 FIXATIVES

B4.1 Acetone

❑ *Reagents*
- Acetone 100%   ↝ AR quality
    ↝ CH$_3$COCH$_3$

❑ *Precaution*
The slide, container, etc., must not be affected by acetone.

Reagents

❑ *Protocol*
Immerse in acetone:
- Cells in culture
- Smears
- Cytocentrifuged cells

10 to 30 min at RT or −20°C   ↪ Better preservation of morphology

❑ *Storage*
The slides can be stored with a desiccant after drying.   ↪ Air-drying

B4.2 Alcohol

❑ *Reagents*   ↪ AR quality
- Distilled water
- Ethyl alcohol 95%   ↪ CH_3CH_2OH
- Methyl alcohol 100%   ↪ CH_3OH

B4.2.1 Ethyl alcohol

❑ *Protocol*
Place the sections or biological samples in the alcohol solution: ↪ Use a pencil to note the references on the slides.
- Alcohol 95%   **15 min to 16 h**   ↪ Duration is not very important.

❑ *Storage*
The slides can be stored in alcohol 70%, or, after drying, in the presence of a desiccant.

B4.2.2 Methyl alcohol

❑ *Precaution*
Always use at 100%.

❑ *Protocol*
Immerse in the alcohol solution
- Alcohol 100%   **15 to 60 min**   ↪ Use a pencil to note the references on the slides.

B4.3 Formol

❑ *Reagents/Solutions*
- Formaldehyde 37 to 40%   ↪ Commercial solution
- 100 mM phosphate buffer or PBS   ↪ *See* Appendix B3.3.

❑ *Precaution*
There are risks to the eyes and the respiratory passages.   ↪ Manipulate under hood.

❑ *Protocol*
Use in 1:10 dilution, in buffer.

Appendix B

❑ *Storage*
Storage is to be avoided after dilution.

B4.4 Paraformaldehyde

❑ *Equipment*
- Hood
- Magnetic stirrer, with heat

❑ *Reagents/Solutions*
- Caustic soda 10 N ↪ AR quality
- Distilled water ↪ See Appendix B2.21.
- Paraformaldehyde in powder ↪ See Appendix B1.1.
  ↪ $HO(CH_2O)_nH$; should be in the form of nonhydrated powder
- 1 M phosphate buffer ↪ See Appendix B3.3.1.

B4.4.1 Paraformaldehyde (40%) ↪ Stock solution

❑ *Precaution*
Manipulate under hood.
❑ *Protocol*
1. In an Erlenmeyer, mix:
   - Distilled water   **100 mL**
   - Paraformaldehyde   **80 g**
2. Allow to hydrate for some minutes, then mix ↪ The solution is milky in appearance.
   the solution with magnetic stirring.
3. Add caustic soda until the solution clears
   - NaOH   **≈0.5 mL**
   - Distilled water   **to 200 mL**
4. Close the receptacle with aluminum foil.
5. Heat and shake.   **20 min** ↪ The solution clears.
                              at 60°C
6. Filter and allow to cool.
❑ *Storage*
- 1 month at 4°C
- Several months in aliquots at –20°C ↪ Do not refreeze.

B4.4.2 Paraformaldehyde (4%) ↪ Working solution

❑ *Precaution*
Manipulate under hood.
❑ *Protocol A*
1. In an Erlenmeyer:
   - 40% paraformaldehyde   **10 mL** ↪ See Appendix B4.4.1.
   - 1 M phosphate buffer   **10 mL** ↪ See Appendix B3.3.1.
   - Distilled water   **50 mL** ↪ See Appendix B1.1.

Reagents

2. Shake
3. Adjust the pH with NaOH   **pH 7.4**
   - Distilled water   **to 100 mL**
4. Filter.

⇝ Filter if necessary.

❏ *Protocol B*

⇝ Dilute the stock solution 1:10 in 100 mM phosphate buffer.

1. In an Erlenmeyer, mix:
   - Paraformaldehyde   **4 g**
   - Distilled water   **50 mL**
2. Shake while heating.   **60°C**
3. Allow to cool, and add:
   - 1 M phosphate buffer   **10 mL**
4. Adjust the pH with NaOH   **pH 7.4**
   - Distilled water   **to 100 mL**
5. Filter and allow to cool.

⇝ *See* Appendix B1.1.

⇝ *See* Appendix B3.3.1.

❏ *Storage*
To be used immediately, or store
At -20°C

⇝ Do not refreeze

B4.4.3   Paraformaldehyde (4%) / Glutaraldehyde (0.05%)

⇝ **Working solution**

❏ *Reagents/Solutions*
- Glutaraldehyde 25 to 75%

⇝ OCH(CH$_2$)$_3$CHO, microscopy quality; the higher the concentration, the more stable the solution.

- 4% paraformaldehyde (PF) in 100 mM phosphate buffer; pH 7.4

⇝ *See* Appendix B4.4.2.

⇝

❏ *Precaution*
Manipulate under hood; do not refreeze.

❏ *Protocol*
- Buffered 4% paraformaldehyde   **200 mL**
- Glutaraldehyde   **133 to 400 µL**

⇝ According to the glutaraldehyde concentration

❏ *Storage*
Use at once.

B5 AUTORADIOGRAPHY

B5.1 Standard Developer

❏ *Reagents/ Solutions*
- Hydroquinone
- Metol
- Potassium bromide
- Sodium carbonate
- Sodium sulfite anhydride
- Sterile water

❏ *Precaution*
Reagent is toxic.

❏ *Working solution*

⇝ AR quality
⇝ $C_6H_4(OH)_2$
⇝ $[OHC_6H_4NH(CH_3)]\cdot 2H_2SO_4$
⇝ KBr
⇝ Na_2CO_3
⇝ Na_2SO_3
⇝ *See* Appendix B1.1.

⇝ Manipulate with gloves.
⇝ Preferably use ready-to-use solutions recommended by the manufacturer of the emulsions and films.

1. Dissolve the following components:
   - Metol   0.5 g
   - Hydroquinone   6 g
   - Sodium sulfite   50 g   ⇝ Anhydride
   - Sodium carbonate   32 g   ⇝ Anhydride
   - Potassium bromide   2 g
   - Running water   to 1 l
2. Filter.

❏ *Storage*
Store at 4°C.

❏ *Protocol*
- Pure   **4 min at 17°C**
- Diluted in water (1:1 $^v/_v$)   **6 min at 17°C**

B5.2 Fixative

❏ *Reagents/Solutions*
- Distilled water
- Sodium thiosulfate   ⇝ $Na_2S_7O_3$

❏ *Working solution*
- Sodium thiosulfate 30%   **60 g**   ⇝ MW = 158.00.
- Distilled water   **200 mL** ⇝ Minimum volume for a staining tray

❏ *Protocol*
- Working solution   **1 to 5 min**

❏ *Storage*
Store for some weeks at room temperature.

B6 IMMUNOCYTOLOGY

B6.1 Blocking Solutions

B6.1.1 Nonspecific sites

❏ *Reagents/Solutions*
Blocking agents:
- Bovine serum albumin (BSA)
- Fish gelatin
- Goat serum
- Ovalbumin
- Skimmed, unsweetened milk
- Triton X-100 0.01%

Buffers:
- 100 mM phosphate buffer; pH 7.4 ⇒ *See* Appendix B3.3.1.
- 50 mM Tris – HCl buffer; 300 mM NaCl; pH 7.6 ⇒ *See* Appendix B3.6.5.

❏ *Precaution*
Do not leave at room temperature.

❏ *Working solution*
Dilute:
- Blocking agents   1 to 2%
- Buffer   to 100 mL

❏ *Storage*
Store at −20°C.

B6.1.2 Endogenous alkaline phosphatases

B6.1.2.1 LEVAMISOLE

⇒ Inhibition is generally not necessary after paraffin embedding. The alkaline phosphatase is deactivated by heat during the embedding step.

❏ *Reagents/Solutions*
- Levamisole ⇒ $C_{11}H_{12}N_2S$ (2,3,5,6-tetrahydro-6-phenyl-imidazole)
- 50 mM Tris – HCl buffer; 100 mM NaCl; 50 mM $MgCl_2$; pH 9.5   ⇒ *See* Appendix B3.6.6.

❏ *Precaution*
None.

❏ *0.1 M stock solution*
Dissolve:
- 0.1 M levamisole   24 mg   ⇒ MW = 240.80.
- Revelation buffer   1 mL

☐ *1 mM working solution*
- Stock solution 10 µL
- Revelation buffer 990 µL

☐ *Storage*
Use immediately.

B6.1.3 Endogenous peroxidases

B6.1.3.1 HYDROGEN PEROXIDE / BUFFER

☐ *Reagents/Solutions*
- Hydrogen peroxide 30% ⇨ 110 vol.
- 50 mM Tris – HCl; 300 mM NaCl; pH 7.6 ⇨ *See* Appendix B3.6.5, or phosphate buffer (*see* Appendix B3.3)

☐ *Precaution*
Avoid contact with the skin. ⇨ Wash abundantly in water.

☐ *Solution for use*
Mix:
- Hydrogen peroxide 3 mL
- Buffer 100 mL

☐ *Storage*
Prepare at time of use.

B6.1.3.2 HYDROGEN PEROXIDE / METHANOL

☐ *Reagents/Solutions*
- 30% hydrogen peroxide ⇨ 110 vol.
- Methanol 100°

☐ *Precaution*
Contact is dangerous.

☐ *Working solution*
Mix:
- Hydrogen peroxide 3 mL
- Methanol 100 mL

☐ *Storage*
Prepare at time of use.

B6.2 Chromogens

B6.2.1 Alkaline phosphatase

B6.2.1.1 NBT – BCIP ⇨ *See* Figure 5.26.
☐ *Reagents/Solutions*
- Dimethylformamide ⇨ $(CH_3)_2NOCH$
- Substrates
  — NBT (Nitroblue tetrazolium) ⇨ $C_{40}H_{30}Cl_2N_{10}O_6$
  — BCIP ⇨ (5-bromo-4-chloro-3-indolyl phosphate)
- Tris – HCl / NaCl / MgCl$_2$ buffer; pH 9.5 ⇨ *See* Appendix B3.6.6.

Reagents

❏ *Precaution*
Avoid contact with the skin.
❏ *Working solution*
Mix:
• NBT	0.34 mg or 4.5 µL	⇝ MW = 817.70. ⇝ 75 mg / mL dimethylformamide
• BCIP	0.18 mg or 3.5 µL	⇝ MW = 433.60. ⇝ 50 mg / mL dimethylformamide
• Revelation buffer	to 1 mL	

❏ *Storage*
Prepare at time of use in darkness. Reagents in commercial solutions are stored at 4°C.

B6.2.1.2 FAST – RED / NAPHTHOL PHOSPHATE

⇝ See Figure 5.27.
⇝ Not to be confused with the rapid red nuclear stain (see Appendix B7.5)

❏ *Reagents/Solutions*
- Dimethylformamide
- Fast – Red
- Naphthol phosphate
- Tris – HCl / NaCl / MgCl$_2$ buffer; pH 9.5

⇝ (CH$_3$)$_2$NOCH
⇝ C$_7$H$_6$N$_2$OCl

⇝ See Appendix B3.6.6.

❏ *Precaution*
Fast–Red is a dangerous reagent when used in powder form.

⇝ Tablet form is preferred.

❏ *Working solution*
1. Dissolve:
• Naphthol substrate	2 mg
• Dimethylformamide	200 µL
2. Add:
• Revelation buffer	9.8 mL
3. Extemporaneously add:
• Fast – Red 1:250	10 mg

⇝ MW = 273.20.

❏ *Storage*
Prepare at time of use.

B6.2.2 Peroxidase

⇝ Inhibition of peroxidase activities

B6.2.2.1 DIAMINOBENZIDINE TETRAHYDROCHLORIDE (DAB)

⇝ See Figure 5.28 (DAB).

❏ *Reagents/Solutions*
- DAB
- 30% hydrogen peroxide
- Tris – HCl / NaCl buffer; pH 7.6

⇝ C$_{12}$H$_{14}$N$_{14}$·HCl
⇝ 110 vol
⇝ See Appendix B3.6.5.

❏ *Precaution*
Reagent is dangerous if inhaled.

⇝ Tablet form is preferable.

❑ *Working solution*
1. Dissolve
   - DAB or 1 tablet   **10 mg**
   - Revelation buffer   **10 mL**
2. Filter the solution.
3. To the filtrate, add:
   - 3% hydrogen peroxide / H_2O **10 µL**

❑ *Storage*
Store at –20°C, in aliquots, without hydrogen peroxide.

↪ Prepare at time of use in darkness.

↪ MW = 360.10
↪ Dissolve tablet of DAB in water.

↪ Commercial solutions (DAB and H_2O_2) can be stored at 4°C.

B6.2.2.2   4-CHLORO-1-NAPHTHOL

❑ *Reagents/Solutions*
- Alcohol 95%
- 30% hydrogen peroxide
- 4-Chloro-1-naphthol
- Tris – HCl / NaCl buffer; pH 7.6

❑ *Precaution*
None.

❑ *Working solution*
1. Dissolve:
   - 4-Chloro-1-naphthol   **50 mg / 230 µL alcohol 95%**
2. Add:
   - Tris – HCl / NaCl buffer   **50 mL**
3. Mix and filter the solution, then add:
   - Hydrogen peroxide   **100 µL**

↪ *See* Figure 5.29.

↪ 110 vol

↪ *See* Appendix B3.6.5.

↪ Prepare at time of use in darkness.

B6.2.2.3   3-AMINO-9-ETHYL CARBAZOLE

❑ *Reagents/Solutions*
- AEC: 3-Amino-9-ethyl carbazole
- 100 mM acetate buffer; pH 5.2
- Dimethylformamide
- 30% hydrogen peroxide

❑ *Precaution*
None.

❑ *Working solution*
1. Dissolve:
   - AEC   **4 mg**
   - Dimethylformamide   **1 mL**
2. Add:
   - Revelation buffer   **14 mL**
3. Filter and add:
   - Hydrogen peroxide   **150 µL**

↪ *See* Figure 5.30 (AEC).

↪ $(CH_3)_2NOCH$
↪ 110 vol

↪ Prepare at time of use in darkness.

B7 STAINS

↪ See Chapter 6 (Table 6.1).

B7.1 Toluidine Blue

❏ *Reagents/Solutions*
- Distilled water
- Sodium tetraborate ↪ $Na_2B_4O_7 \cdot 12H_2O$
- Toluidine blue ↪ $C_{15}H_{16}N_3 \cdot SCl$

❏ *Precaution*
Filter the solution before use.

❏ *Stock solution*
1. Mix and dissolve at room temperature:
    - Toluidine blue   **1 g**   ↪ MW = 305.80.
    - Sodium tetraborate   **1 g**   ↪ MW = 381.37.
    - Distilled water   **100 mL**
2. Filter.

❏ *Working solution*
Dilute:
- Stock solution   **1 vol**
- Distilled water   **50 vol**

❏ *Storage*
Store at room temperature for some months.

B7.2 Cresyl Violet

❏ *Reagents/Solutions*
- Cresyl violet
- Distilled water

❏ *Precaution*
Filter the solution before use.

❏ *Working solution*
Dissolve:
- Cresyl violet   **1 g**
- Distilled water   **100 mL**

❏ *Storage*
Store at room temperature for some weeks.

B7.3 Eosin

❏ *Reagents/Solutions*
- Distilled water
- Eosin

❏ *Precaution*
None.

❏ *Working solution*
Dissolve:
- Eosin **5 g**
- Distilled water **100 mL**

❏ *Storage*
Store at room temperature in an opaque container for several months.

B7.4 Harris's Hematoxylin

❏ *Reagents/Solutions*
- Acetic acid ⇀ $C_2H_4O_2$
- Alum ⇀ Aluminum potassium sulfate $(AlK(SO_4)_2 \cdot 12H_2O)$
- Distilled water
- Ethyl alcohol
- Harris's hematoxylin ⇀ $C_{16}H_{14}O_6$
- Yellow mercury oxide

❏ *Precautions*
Mercury oxide is toxic.
Filter the solution before use.
The solution matures with time. ⇀ This brings about an increase in its dyeing properties.

❏ *Working solution*
1. Dissolve:
   - Hematoxylin **5 g** ⇀ MW = 302.30.
   - Ethyl alcohol **50 mL**
2. Dissolve while warm:
   - Alum **100 g** ⇀ MW = 474.39.
   - Distilled water **1 l**
3. Mix 1 and 2
4. Heat to boiling point and add:
   - Mercury oxide **2.5 g**
5. Withdraw rapidly from the heat and add:
   - Acetic acid **20 mL** ⇀ MW = 60.05.

❏ *Storage*
Store at room temperature, in an opaque container, for some months.

B7.5 Rapid Nuclear Red

↪ Not to be confused with Fast – Red.

❑ *Reagents/Solutions*
- Aluminum sulfate   ↪ $Al_2(SO_4)_3$
- Distilled water
- Rapid nuclear red   ↪ $C_{14}H_9NO_7S$

❑ *Precaution*
Filter the solution before use.

❑ *Working solution*
1. Prepare 1 L of aluminum sulfate at 5%:
   - Aluminum sulfate   **50 g**   ↪ MW = 342.1.
   - Distilled water   **1 L**
2. Heat to boiling point.
3. Withdraw from the heat, and add:
   - Nuclear red   **1 g**   ↪ MW = 335.3.

❑ *Storage*
Store at 4°C.

B7.6 Methyl Green

❑ *Reagents/Solutions*
- Distilled water
- Methyl green   ↪ $C_{27}H_{35}N_3BrCl \cdot ZnCl_2$

❑ *Precaution*
Filter before use.

❑ *Working solution*
Dissolve:
- Methyl green   **1 g**   ↪ MW = 653.2.
- Distilled water   **100 mL**

❑ *Storage*
Store at room temperature.

B8 MOUNTING MEDIA

B8.1 Aqueous

Aqueous mounts are used where the reaction precipitate is soluble in alcohol.
- Buffered glycerine   ↪ *See* Appendix B8.1.1.
- Commercial media
- Moviol   ↪ *See* Appendix B8.1.2.

B8.1.1 Buffered glycerine

❑ *Reagents/Solutions*
- Glycerol
- 100 mM phosphate buffer, or PBS ↦ *See* Appendix B3.3.

❑ *Precaution*
None.

❑ *Preparation*
- Glycerol   1 vol
- Buffer   1 vol

❑ *Storage*
Store at 4°C for some days.

B8.1.2 Moviol

❑ *Reagents/Solutions*
- Distilled water ↦ AR quality
- Glycerol anhydride ↦ $C_3H_8O_3$
- Moviol 4 – 88
- 100 mM Tris – HCl buffer; pH 8.5 ↦ *See* Appendix B3.6.1.

❑ *Precaution*
None.

❑ *Preparation*
1. Mix:
   - Moviol 4 – 88   2.4 g
   - Glycerol   4.8 mL ↦ MW = 92.10.
2. Add:
   - Distilled water 6 mL
3. Mix and leave.   2 to 48 h at 60°C
4. Add:
   - Buffer   12 mL
5. Incubate.   Overnight at 60°C
6. Mix, centrifuge. 5000 g for 15 min

❑ *Storage*
Store at –20°C in aliquots.

B8.2 Permanent

Permanent mounting media are used after dehydration and passage through xylene (or a similar chemical).

Reagents

❑ *Reagents*
• Commercial media
❑ *Precautions*
Toxic on contact or inhalation
Polymerization on drying
❑ *Storage*
Store at room temperature, in stoppered receptacles.

↪ Aromatic solvents, or substitutes

↪ Fluidification with xylene is possible.

B8.3 For Fluorescence

Anti-fading additives that are present in some mounting media facilitate the conservation and amplification of fluorescence.
Some aqueous mounts can also be used.

❑ *Reagents/Solutions*
• Commercial media
• Revelation buffer
❑ *Precaution*
None.
❑ *Storage*
Store at 4°C.

↪ For example, Moviol (*see* Appendix B8.1.2).

↪ *See* Appendices B3.6.1 and B3.6.5.

Glossary

A

Acetylation ⇝ Transformation of the amino group, $-NH_2$ (reactive), in a protein into a substituted amide group, $-NH-CO-CH_3$ (neutral), by the action of acetic anhydride, $CH_3-CO-CH_3$, in a triethanolamine buffer, pH 8. This reaction is effective in reducing background noise levels, although it also acts on the nitrogenous bases of nucleic acids, and can cause a reduction in the hybridization signal.

Acid stain ⇝ An anionic stain. It carries negative charges, such as the carboxyl group, and stains positively charged substances.

Adenine ⇝ A puric base that is found in nucleosides, nucleotides, coenzymes, and nucleic acids. It establishes two hydrogen bonds with thymine and deoxyuracil in double-stranded nucleic acids.

Alkaline phosphatase ⇝ A phosphatase extracted from calf intestine (MW = 80 kDa). Its optimum pH is alkaline.

Amino acid ⇝ The elementary unit in proteins. Each one contains an amino group, $-NH_2$, and an organic acid group, $-COOH$. The amino acid chains that make up the proteins are called polypeptide chains.

Ångström ⇝ $1 \text{ Å} = 10^{-1}$ nm (or 10^{-10} m).

Antibody ⇝ A glycoprotein produced as a response to the introduction of an antigen. It can combine with the antigen that induced its production. *See also* Immunoglobulin.

Antigen ⇝ Any substance that is foreign to a receiving organism is said to be immunogenic when it induces an immune response (which can engender the production of antibodies).

Antigenic determinant ⇝ *See* Epitope.

Anti-sense probe ⇝ *See* Probe.

Antiserum ⇝ Serum containing immunoglobulins.

Artifact ⇝ Background noise, nonspecific reactions.

ATPase ⇝ Adenosine triphosphatase (an enzyme that catalyzes the dephosphorylation of ATP).

Autoclave ⇝ A device that is used to sterilize objects using pressurized steam.

Autoradiographic activity ⇝ An expression of the number of disintegrations per second that take place in a radioactive source. The unit is the Becquerel, which corresponds to one disintegration per second; the previous unit was the Curie (1 Bq = 2.7×10^{-11} Curie)

Glossary

Autoradiographic efficacy	⇝ The number of silver grains obtained by β^- radiation.
Autoradiographic resolution	⇝ Distance between the location of a silver grain and the source of the radioactive emission (a function of the energy of the isotope and the autoradiographic system).
Autoradiography / radioautography	⇝ Method for visualizing hybrids containing isotopes.
Avidin	⇝ A glycoprotein (MW = 16,400) produced in the uterus of the chicken. It contains four subunits, each of which binds a molecule of biotin with a very high affinity: kDa 10^{-15}.

B

Background	⇝ Nonspecific signal (artifact).
Base	⇝ There are five bases whose combination expresses the genetic code: adenine, guanine, cytosine, and thymine or uracil. They are divided into two groups: purines and pyrimidines (*see* Section 4.2).
Base pair	⇝ A parameter that is used for expressing the length of a double strand of DNA. A base pair consists of an association of two nucleotides, one on each of the two complementary strands of DNA.
Basic stain	⇝ A cationic stain. It carries positive charges that fix to negatively charged cell structures.
Bright-field microscope	⇝ A type of microscope whose object is directly illuminated in an intense way and forms an image on a brighter background.

C

Cellular clone	⇝ All the cells that are produced by the division of a given mother cell (genetically identical cells).
Chelator	⇝ Agent (e.g., EDTA) that captures Ca^{2+} ions. It blocks enzymatic action.
Chemography	⇝ Chemical effect of the biological preparation on the photographic emulsion which causes background.
Chromogen	⇝ Substrate modified by an enzyme to produce a colored substance.

Chromosome	⇝ The name comes from the Greek *khroma*, which means "color." Its structure contains most or all of the cellular DNA, and thus the genetic information.
Cloning vector	⇝ A DNA molecule that is capable of replicating itself (replicon). It is used to transport a piece of foreign DNA (e.g., a gene) into a target cell. This vector can be a plasmid, a phage, etc.
Codon	⇝ Sequence of three bases (triplet) in the mRNA that codes for an amino acid. It can also signal the initiation or the end of the translation process.
Complementary DNA (cDNA)	⇝ A DNA copy of an RNA molecule, generally messenger RNA.
Counterstaining	⇝ Secondary staining that makes it possible to obtain a different contrast from that of the hybridization signal.
Cryosection	⇝ Section cut from frozen tissue.
Cytocentrifugation	⇝ Centrifugation of cell suspensions on slides.
Cytosine	⇝ A pyrimidic base that is a component of nucleosides, nucleotides, and nucleic acids.

D

Dark-field microscope	⇝ A type of microscope whose object is illuminated laterally. Only objects that reflect light are visible, and the tissue background remains black.
Dehydration	⇝ Successive baths in increasing concentrations of a solvent (e.g., alcohol, acetone, etc.), which is substituted for water in the tissue.
Denaturation	⇝ Separation of the two strands of the double helix of DNA, either by heat or through the action of other chemical agents (change of configuration of the nucleic acids). It can also be applied to certain RNAs.
Denhardt's solution	⇝ Mixture of polymers that reduces background.
Deoxynucleoside	⇝ Molecule consisting of a combination of a base and a deoxyribose.
Deoxynucleotide	⇝ Molecule consisting of a combination of a base, a deoxyribose, and one or more phosphates.
Deoxyribonucleic acid (DNA)	⇝ Nucleic acid that is found in the nucleus of the cell (and carries genetic information in all cells). The DNA molecule is a polynucleotide that is composed essentially of deoxyribonucleotides linked by phosphodiester bonds.

Glossary

Deoxyribose	↝ A ribose that has lost an atom of oxygen at 2' (*See* Figure 4.6).
Deproteinization	↝ Enzymatic treatment that partly eliminates proteins, and, more particularly, those that are associated with nucleic acids.
Dextran sulfate	↝ Substance that increases the efficiency of the hybridization process by concentrating the probe. Useful in the case of cDNA probes.
DNA ligase	↝ An enzyme combining two fragments of DNA by the formation of a new phosphodiester bond.
DNA polymerase	↝ An enzyme that synthesizes a new DNA molecule from a DNA matrix.
DNase	↝ An enzyme that destroys DNA in a specific way.
Duplication	↝ The process by which the DNA molecule is able to make an exact copy of itself.

E

ELISA test	↝ "Enzyme-linked immunosorbent assay": technique that is used to detect and quantify specific antibodies and antigens.
Endogen	↝ A molecule that is found in a biological sample.
Energy of emission	↝ Energy liberated by the disintegration of an isotope (the energy is measured in MeV, i.e., megaelectronvolts).
Enzyme	↝ A protein that catalyzes a specific chemical reaction. The name of an enzyme generally specifies its function. The addition of the suffix–*ase* to the name of the type of molecule on which the enzyme acts means that it is broken down by this enzyme (e.g., ribonuclease cuts RNA molecules; proteases hydrolyze proteins).
Epipolarization microscope	↝ Microscope in which polarized light can be used to reinforce the contrast observed with certain transparent or opaque reflecting objects.
Epitope	↝ Also termed "antigenic determinant": the part of the antigen that is recognized by an antibody (an epitope is a structure that can combine with a single antibody molecule). An antigen can have several epitopes, each one corresponding to a sequence of three to six contiguous or noncontiguous amino acids.

Glossary

Exogen ↪ A substance introduced by contamination from outside (recipients, etc.).

Exon ↪ A translated (or coding) sequence of DNA.

Expression vector ↪ A particular cloning vector for expressing recombinant genes in host cells. The recombinant gene is transcribed, and the protein for which it codes is synthesized.

F

Fab ↪ Antigen-binding fragment of immunoglobulin (MW = 45 kDa) obtained by cutting an antibody with an enzyme (papain). The enzymatic cut is situated upstream from the disulfur bonds that link the two heavy chains of the immunoglobulin. The fragment represents $2/3$ of the native molecule.

Fc fragment ↪ A fragment (MW = 50 kDa) which represents $1/3$ of the native molecule. It has no antibody activity, but is easily crystallized; hence its name.

Fixation ↪ A process whereby a fixing solution is used to fix the cell metabolism and deactivate endogenous RNases, while conserving cell morphology and the integrity of the nucleic acids.

Fluorescence ↪ A technique using stains called fluorochromes which are visualized under a fluorescence microscope.

Fluorescence microscope ↪ A type of microscope in which the sample is exposed to ultraviolet light of specific wavelength. This excites a fluorochrome, which, in response to the excitation, emits light in the visible spectrum. The sample can usually be stained.

Fluorochrome ↪ Molecule that contains conjugated double bonds, and that, as a result of stimulation by a photon, responds by emitting another photon whose wavelength is longer than that of the exciting photon. This structural property is responsible for the phenomenon of fluorescence under ultraviolet excitation.

Formamide ↪ A molecule that favors the denaturation of nucleic acids. The addition of 1% formamide to the hybridization solution lowers the hybridization temperature by 0.65°C for DNA and by 0.72°C for RNA.

Glossary

G

Gene — A segment of DNA involved in producing a polypeptide chain. It includes regions preceding and following the coding region (exons) and introns.

Genetic engineering — In general terms, the techniques used to deliberately modify genetic information about an organism through changes in its genomic nucleic acid.

Genome — The entire set of genes to be found in a eukaryotic cell, bacterium, yeast, or virus.

Genotype — Representation of the totality of genetic information for all the genes of a given organism.

Guanine — A puric base that is found in nucleosides, nucleotides, and nucleic acids.

H

Half-life of a radioelement — Time required for radioactivity to decrease by half.

Hapten — A nonimmunogenic molecule, which is nonetheless able to induce the production of antibodies directed against itself when it is coupled to a macromolecular carrier.

Haptenization — Labeling of nuclear probes using haptenes introduced either by a chemical reaction or by enzymatic action.

Histones — Proteins that are rich in arginine and lysine, associated with DNA in eukaryotic cells and chromosomes.

Horseradish peroxidase — An enzyme extracted from horseradish, a plant of the Cruciferae family.

Hybridization of nucleic acids — The formation of hybrid molecules of double-stranded DNA, DNA – RNA, or RNA – RNA. Insofar as the sequences are complementary, stable hybrids will be formed.

Hydrophilic — Said of a polar substance that has a strong affinity for water, or that dissolves easily in water.

Hydrophobic — Said of a nonpolar substance that has no affinity for water, or that does not easily dissolve in water.

I

Immunocytochemistry	⇝ A science that provides cytological evidence for antigenic constituents by means of an immunochemical reaction.
Immunofluorescence	⇝ A technique using stains called fluorochromes which, when exposed to ultraviolet light, become fluorescent; i.e., they emit visible light.
Immunoglobulin	⇝ *See* Antibody.
Immunoglobulin G (IgG)	⇝ A class of immunoglobulins that are predominant in serum (> 85% of the immunoglobulins in serum) (MW = 150 kDa).
Immunoglobulin M (IgM)	⇝ A class of serous immunoglobulins (MW = 900 kDa) formed of heavy chains and light chains such as IgGs. The molecules are in the shape of a five-pointed star with a central ring.
Immunohistochemistry	⇝ Immunochemistry + histology.
Inclusion	⇝ A step that consists of replacing tissue water with a hard substance (e.g., paraffin or resin).
Intron	⇝ Segment of DNA that is transcribed, but excised during splicing, leading to the maturation of mRNA.
Isotopes	⇝ Atoms of a given element that have different atomic masses, although their chemical properties are very similar.

M

Melting temperature (Tm)	⇝ The temperature at which 50% of double-stranded DNA separates out into individual chains. It depends on the C and G content of the DNA, the Na^+ concentration, and the length of the strands (*See* Section 4.2.3.1).
mer	⇝ Unit of nucleic acid: 1 mer corresponds to 1 nucleotide or 1 base pair.
Messenger ribonucleic acid (mRNA)	⇝ Single-stranded RNA synthesized from a DNA matrix during transcription. It links to ribosomes and carries messages used for protein synthesis.
Micrometer	⇝ $1 \mu m = 10^{-3}$ mm (or 10^{-6} m).
Microscopic resolution	⇝ The ability of a microscope to distinguish between two objects.
Monoclonal antibody	⇝ An antibody that is specific to an epitope produced by a cell culture resulting from the fusion of a tumoral cell and a cell that produces antibodies.

Glossary

N

Nanometer ↪ 1 nm = 10^{-3} µm (or 10^{-9} m).

Nick translation ↪ A method for labeling DNA, also a method for repairing breaks (*See* Section 1.3.2).

Nonimmune serum ↪ Serum from an animal of the same species.

Nonspecificity ↪ Nonhomogeneous (probe – noncomplementary nucleic acid), or heterogeneous (protein – nucleic acid) matching.

Northern blot technique ↪ Hybridization of RNA on a membrane.

Nucleic acids ↪ Macromolecules, the vehicles for genetic information.

Nucleolus ↪ An organelle situated in the nuclear matrix and not limited by a membrane, where ribosomal RNA synthesis and the assembly of ribosomal subunits take place.

Nucleoside ↪ A molecule that is a combination of a puric or pyrimidic base and a sugar (ribose) (*see* Section 4.2.1).

Nucleosome ↪ A molecule of DNA associated with basic proteins (histones) in the eukaryotic nuclear matrix. Histones participate in the formation of chromatins, which are made up of flexible chains of repeated units, namely, the nucleosomes.

Nucleotide ↪ A nucleoside + one or more phosphates (*see* Section 4.2.1).

Nucleotide triphosphate ↪ A puric or pyrimidic base + a sugar (ribose or deoxyribose) + three phosphate groups (*see* Section 4.2.1).

O

Oligonucleotide ↪ From the Greek *oligoi*, meaning "few." Oligonucleotides are short fragments of DNA, with 15 to 60 nucleotides.

Oligonucleotide probe ↪ In cases where the sequence is known, it is possible to make a single-stranded probe using synthetic oligonucleotides ($\approx$ 30 nucleotides) whose sequences are complementary to those of the nucleic acids being sought.

P

Peptide ↝ A short chain of amino acids.

Phase-contrast microscope ↝ A type of microscope that converts small differences in refractive index and cell density into observable differences in luminous intensity.

Phenotype ↝ A set of characteristics of an organism resulting from interaction of the genome and the environment.

Phosphatase ↝ An enzyme that hydrolyzes phosphate groups in molecules.

Phosphorylation ↝ The addition of a phosphate group by the action of an enzyme.

Plasmid ↝ A molecule of DNA into which it is possible to introduce a DNA insert. It is used as a cloning vector.

Poly A ↝ A repetitive polynucleotide sequence that is attached by the action of a ligase to the 3' end of a transcribed mRNA molecule. This is characteristic of mRNA.

Polyclonal antibody ↝ All the immunoglobulins contained in the serum from an animal immunized with a given antigen. Polyclonal antibodies recognize several epitopes or antigenic determinants.

Polymerase ↝ A synthesization enzyme for DNA or RNA.

Polymerase chain reaction (PCR) ↝ A reaction that takes place in repeated cycles (denaturation, hybridization, elongation), which makes possible the amplification of a given sequence of DNA.

Polynucleotide ↝ A chain formed by a combination of several nucleotides (RNA is a polynucleotide).

Polynucleotide kinase ↝ The kinase enzyme (which is extracted from calf thymus) that is used in the radioactive labeling of oligonucleotide probes by phosphorylating the 5' end (phosphate group in γ position).

Postfixation ↝ Second fixation step.

Precipitation reaction ↝ A reaction that causes the desolubilization of a nucleic acid, using alcohol and a salt.

Prehybridization ↝ Incubation of sections in the hybridization solution without the probe.

Preimmune serum ↝ Serum from an animal before immunization.

Pre-messenger RNA ↝ Single-stranded RNA

Primary antibody ↝ The first antibody used in the specific immunological reaction of the antigen being sought.

Glossary

Probe ↪ A fragment of nucleic acid (DNA or RNA) whose nucleotide sequence is complementary to that of the nucleic acid being sought, which is immobilized in the preparation (target).

Anti-sense probe ↪ A complementary sequence to that of the target: it hybridizes specifically with the complementary sequence that is immobilized *in situ* (target).

Sense probe ↪ A sense probe is a homologous copy to that of the target. It serves as a negative control.

Prokaryotic cell ↪ Cell that has no nucleus (blue-green algae, bacteria).

Promoter ↪ A sequence of a sense strand of DNA situated upstream from a gene, to which polymerase RNA is linked before the start of the transcription process.

Protein ↪ A macromolecule composed of amino acids.

Proteinase ↪ Enzyme that hydrolyzes proteins into amino acids.

Puric bases ↪ The puric bases are guanine and adenine (*see* Section 4.2.2), whose chemical structures resemble those of the pyrimidines, combined with a pentagonal structure (pyrimidic cycle + imidazol cycle).

Purine ↪ A basic nitrogenous molecule composed of two aromatic nuclei, found in nucleic acids and other cell components. It is made up principally of the bases adenine and guanine.

Pyrimidic bases ↪ The pyrimidic bases are cytosine and thymine (DNA) or uracil (RNA), whose chemical structures are hexagonal.

Pyrimidine ↪ A basic nitrogenous molecule composed of an aromatic nucleus, found in nucleic acids and other cell components. It is made up principally of the bases cytosine, uracil and thymine.

R

Radioactivity ↪ A property possessed by certain elements of transforming themselves by disintegration into other elements, with the emission of different forms of radiation.

Random priming ↪ A method for random labeling by duplication of all the DNA, starting with nonspecific primers (*see* Section 1.3.1).

Glossary

Recombinant DNA technology ↬ Techniques used in genetic engineering for the identification and isolation of specific genes, their insertion into a vector such as a plasmid to form recombinant DNA, and the production of large quantities of the gene and its product.

Renaturation ↬ Rematching of complementary nucleic acids.

Restriction enzyme ↬ A catalytic protein that cuts DNA at a specific site. Such enzymes are indispensable to the manipulation of DNA *in vitro*.

Reverse transcriptase ↬ An RNA-dependent DNA polymerase. It synthesizes complementary DNA using an RNA matrix.

Ribonuclease (RNase) ↬ *See* RNase.

Ribonucleic acid (RNA) ↬ A polynucleotide composed of ribonucleotides linked by phosphodiester bonds. There is messenger RNA, transfer RNA, ribosomic RNA, and viral RNA.

Ribose ↬ A ribonucleic acid (sugar in RNA).

Ribosomal ribonucleic acid (rRNA) ↬ A component of ribosomes. Several single-stranded rRNA molecules of different sizes contribute to the structure of ribosomes, and are also directly involved in protein synthesis.

RNA polymerase ↬ Enzyme that catalyzes the synthesis of RNA from a DNA matrix. RNA polymerase recognizes the errors that result from incorrect matching. Thanks to its exonucleasic activity, it removes erroneous nucleotides, for which it substitutes the appropriate nucleotides.

RNase ↬ An enzyme that breaks down only single-stranded RNA. RNase treatment carried out after hybridization makes it possible to reduce background noise.

RNase-free conditions ↬ Experimental conditions in which all contamination by exogenous ribonucleases is eliminated, to preserve mRNA.

S

Saline concentration ↬ Ionic strength, *see* Salinity.

Salinity ↬ The concentration of Na ions, which influences the stability of the hybrids. The hybridization speed increases with the concentration of salts.

Secondary antibody ↬ The second antibody used in indirect immunological reactions, whether labeled or not, directed against the animal species that was used to produce the IgGs of the primary antibody.

Sense probe ↬ *See* Probe.

Glossary

Sensitivity — The smallest quantity of nucleic acid-target that can be detected in a cell or tissue, or the number of molecules that can be detected by a given *in situ* hybridization reaction.

Southern blot technique — Detection of specific fragments of DNA transferred to a membrane by hybridization with a labeled probe.

Specific activity — The specific activity of a probe results from its labeling; i.e., the number of isotopes or antigens incorporated, by comparison with the mass or the concentration, of the probe.

Specificity — Total complementarity of matching between two molecules of nucleic acid.

Spliceosome — A site where the splicing of the mRNA precursor takes place.

Splicing of RNA — A process during which introns are cut out of the primary transcripts of RNA in the nucleus during the formation of mRNA.

Stability — Condition between two molecules of nucleic acid. It depends on their nature, and there are three types of duplex that can be formed: DNA – DNA, DNA – RNA, RNA – RNA, in increasing order of stability.

Sterilization — A process whereby all living cells and viruses are either destroyed or eliminated from an object or a habitat.

Streptavidin — A protein of bacterial origin that has a high affinity for four biotin molecules and a very weak charge, and that causes little background. Its characteristics are similar to those of avidin.

Stringency — A parameter that is used to express the efficacy of hybridization and washing conditions (depending on the concentrations of salt and formamide, and the temperature). A low level of stringency favors nonspecific matchings, whereas too high a level gives rise to a specific signal of low intensity.

Substrate — A substance that is acted on by an enzyme.

T

Target — The nucleic sequence that is being sought within the cell.

TdT terminal deoxytransferase — An enzyme that is used for labeling oligonucleotide probes by elongation of the 3' end, provided that this is hydroxylated (free OH). It can polymerize NTP and dNTP.

Glossary

Thymine	⇝ A pyrimidic base that is found in nucleosides, nucleotides, and DNA.
Transcriptase	⇝ Enzyme that catalyzes transcription. In RNA viruses it is an RNA-dependent RNA polymerase which is used to make mRNA from RNA genomes.
Transcription	⇝ A process in the course of which single-stranded RNA is synthesized using a DNA matrix.
Transfer ribonucleic acid (tRNA)	⇝ A short RNA chain that transports amino acids during the synthesis of proteins.
Transferase	⇝ A type of enzyme that catalyzes transfers of methyl, carboxyl, and amino groups, etc.
Transgenic	⇝ An animal or plant whose genome contains new genetic information in a stable manner because of the acquisition of foreign DNA.

U

Ultraviolet light	⇝ Radiation of short wavelength (10 to 400 nm) and high energy.
Uracil	⇝ A pyrimidic base that is found in nucleosides, nucleotides, and RNA.

V

Viral ribonucleic acid	⇝ RNA found in viruses.

Index

A

Acetic anhydride, 95, 102
Acetone, 67
Acetylation, 94, 102, 215
Acid stains, 192
Adenine, 116, 117
Agarose gels, 24, 57
Alcohol, 67, 240
Aldehyde fixatives, 68
Alkaline phosphatase, 205, 247
Aminomethylcoumarin, 160,163
Ammonium acetate, 54
Amplification, 8, 9, 19, 32, 33, 36, 37, 48
    direct visualization of, 183
    fluorochrome, 183
    PCR, 8, 32, 36
    tyramide, 180
Anti-biotin, 17, 57, 172
Antibodies, 157, 165, 172
Antigen
    photoactive, 45, 46, 47
    tyramide, 182, 185
Antigenic probes, *see* Probes, antigenic
Antioxidant, 129
Anti-sense probe, 10
Aqueous mounts, 196
Artifactual labeling, 147
Autoradiography, 12, 143
    developing, 241
    emulsion, choice of, 151

B

Background noise, 178
Biopsies, 65
Biotin, 17, 18, 19, 26, 27, 28, 30, 31, 40, 41, 42, 43, 50, 53, 57, 156, 166
    conjugates, 166
    detection, 57
    formation, 173
    label, 171
    tyramide–, 181
Bipotassium phosphate, 95
Blocking buffers, 172
Blue precipitate, 179
Bovine serum albumin (BSA), 95
Bright field light microscopy, 200, 201
BSA, *see* Bovine serum albumin

C

Calcium chloride, 95
Carmine, 191
cDNA probe
    antigenic, 228
    radioactive, 225
Cell(s)
    cytocentrifuged, 99
    detachment of, 73, 234
    preparation of, 234
    spins, 68
    in suspension, 70, 72
Cellular resolution, 151
Cellular smears, 65
Chromatography, 26
Chromogens, 158
Cobalt chloride, 41, 50
Coding region, sequence comprising, 9
Colloidal gold, 163, 206
Column, passage through, 27, 31, 53, 55
Complementarity, 111
Confocal microscope, 216
Controls and problems, 207–220
    controls, 217–218
        hybridization, washing, 218
        labeling, 217
        probes, 217
        revelation, 218
        tissue, 217
        variation of signal, 218
        verification by other techniques, 218
    recapitulative table, 219–220
    sensitivity/specificity, 214–216
        hybridization, 215
        observation, 216
        pretreatments, 215
        probes, 214
        revelation, 216
        samples, 214
        washing, 216
    signal/background ratio, 211–213
        hybridization, 212–213
        observation, 213
        pretreatments, 212
        probes, 211
        revelation, 213
        samples, 212
        washing, 213
Counterstaining and modes of observation, 187–206
    counterstaining, 191–194
        choice of stain, 192
        classification of stains, 191–192
        protocols, 193–194
    modes of observations, 197–205
        bright field, 201
        dark field, 202
        epipolarization, 203
        fluorescence, 204–205
        light microscope, 197–200
    mounting media, 194–197
        aqueous, 196
        for fluorescence, 197
        permanent, 195–196
    observation of signal, 205–206
        choice of mode of observation, 206
        enzymatic labels, 205
Cresyl violet, 192, 193

Index

cRNA probe, 121
    antigenic, 245
    protocol for, 135
    radioactive, 227, 239
Cross-linking fixatives, 67
Cryomicrotome, 75, 78
Cryoprotection, 74, 76, 232
Cytocentrifugation, 70, 73
Cytosine, 116, 117, 119

D

Dark field microscopy, 202
Dehydration, 74, 80, 102, 126, 235
    in ethanol baths, 84
    mounting after, 242
Denaturation, 24, 25, 26, 27, 32, 33, 35, 36, 37, 39, 46, 94, 112, 219
    of probe, 59
    step, 93
    of target nucleic acid, 104
Denhardt's solution, 129
Deoxyribonucleic acid (DNA), 7, 95, 115, 116
    denaturing of target, 236
    digestion of, 52
    double-stranded, 7, 20, 46
    duplication of, 22
    genomic, 32, 36
    linearized, 34, 38
    plasmid, 31
    polymerase, 27
    probe(s), 59
        double-stranded, 9, 22, 26, 28, 43
        single-stranded, 9, 36
        stability of, 32
    single-stranded, 7, 8, 9, 12, 22, 26, 36, 213
    types of molecules interacting with, 46
    uniform labeling of, 23
Deoxyribonucleotide(s)
    radioactive, 28
    triphosphate, 15
Deoxyribose, 114, 117
Deproteinization, 93, 215, 219, 235
Dewaxing, 88, 233
Dextran sulfate, 95, 239
Diaminobenzidine tetrahydrochloride, 178
Diester bond, 16, 23, 28
Digitonin, 95, 98
Digoxigenin, 17, 18, 26, 27, 28, 30, 31, 40, 41, 42, 43, 50, 53, 57, 156
    detection, 58
    tyramide conjugated to, 185
Dimethylformamide, 176
Dimethylsulfoxide (DMSO), 76
Direct reaction, 171, 174
Direct visualization
    of amplification, 183
    technique, 167
Disodium phosphate, 95

Distilled water, 154, 242
Dithiothreitol (DTT), 25, 127, 131, 240
DMSO, *see* Dimethylsulfoxide
DNA, *see* Deoxyribonucleic acid
DNase-free reagents, 34, 38, 54, 55
Dot blot verification, 218
Double-stranded DNA, 7, 20
    incubation of photoreactive reagent with, 46
    probes, 9
Drying, 81, 126, 155
DTT, *see* Dithiothreitol
Duplication, 118

E

EDTA, *see* Ethylene diamine tetra-acetic acid
Electrophoresis tank, 50, 57
Elongation, 32, 35, 36, 39
Embedding, 84, 233
Emission energy, 13, 14
Emulsion
    -coated slides, drying of, 155
    dilution of, 154
    liquefaction of, 152, 241
    liquid, 146
    photographic, 151
Energy
    emission, 13, 14
    of radiation, 146
    radioactive nucleotide of high, 44
Enlargements, 199, 202
Enzymatic action, 44
Enzymatic cutting, 23
Enzymatic digestion, 30
Enzymatic labels, 158, 175
Enzymatic treatment, destruction of target by, 217
Enzymes, 100, 206
    deoxynucleotide terminal transferase, 43
    inhibition of by EDTA, 52
    Klenow, 22, 23, 24, 25, 26
    tyramide, 182
Eosin, 192, 194
Epi-fluorescence, 204
Epipolarization, 200, 203
Escherichia coli, 159
Ethanol precipitation, 26, 27, 39, 54
Ethidium bromide, 46
Ethylene diamine tetra-acetic acid (EDTA), 25, 52, 95
Exonuclease activity, 23, 28, 29
Exposure time, 15, 16

F

Fast–Red, 177
Ficoll 400, 95
Film
    developer characteristics, 149
    double-coated, 148
    positioning of, 148

single-coated, 147
support, 143
FITC, *see* Fluorescein isothiocyanate
Fixation, 212, 233
   conditions, 67
   freezing after, 77
   by immersion, 70, 72, 82
   by perfusion, 70, 71
   protocols, 71
Fixative(s)
   aldehyde, 68
   concentration of, 67
   cross-linking, 67
   precipitating, 67, 69
Fluorescein, 17, 19, 28, 30, 31, 40, 156
   isothiocyanate (FITC), 160, 161
   tyramide conjugated to, 185
   visualization of, 58
Fluorescence, 143
   endogenous, 167
   light microscopy, 200, 204
   transmitted, 204
Fluorescent labels, 158
Fluorochrome(s), 143, 206
   amplification of, 183
   tyramide, 182
Formaldehyde, 67, 68
Formamide, 95, 122, 127
   concentration, 220
   deionized, 96, 128, 129
Freezing, 74, 76, 232
   after fixation, 77
   rapid, 79
Frozen tissue, 74

G

Gel drying, 57
Genomic DNA, 32
Giemsa stain, 192
Glutaraldehyde, 67, 68, 95
Goat serum, 172
Gold
   colloidal, 163, 206
   latensification of, 174
Guanine, 116, 117

H

Half-life, 13, 14
Harris' hematoxylin, 193
Heating block, 105, 106
Hematoxylin, 191, 193, 194
Heparin, 127, 128
Hexanucleotides, 22, 25, 26, 27
Hybrid
   matched, 123
   mismatched, 123

Hybrid(s), *see also* Revelation of hybrids
   antigenic, 156
   formation of, 120
   loss of stability of, 211
   nature of, 112
   radioactive, 143
   unstable, 113
Hybridization, 35, 39, 60, 107–138
   before revelation, 137–138
      antigenic label, 138
      radioactive label, 138
   buffer, 130, 236
   composition of hybridization medium, 129
   efficacy of, 97
   equipment/reagents/solutions, 127–128
      equipment, 127
      reagents, 127
      solutions, 128
   medium, 103
   opposite of, 112
   parameters, 114–123
      complementarity of bases, 119–120
      homology of sequence, 123
      hybridization buffer, 122
      hybridization temperature, 120–122
      length of probes, 122
      Na^+ concentration, 122
      structure of nucleic acids, 114–119
      type of label used, 123
      type of probes, 122
   posthybridization treatments, 123–125
      purpose of washing, 123–124
      stringency conditions, 124–125
   principle of *in situ* hybridization, 111–113
   process, sterility of solutions for, 96
   protocols, 130–137
      cDNA probe, 130–133
      cRNA probe, 135–137
      oligonucleotide probe, 133–135
   solution, 134
   summary of different steps, 126
   temperature, 113
Hydrocarbons, aliphatic, 83
Hydrophobic medium, 87

I

Image analysis, 216, 220
Immersion, fixation by, 70, 71, 72, 82
Immunocytology, 218, 246
Immunohistology, 12, 143, 157, 247
Impregnation, 84
Incubation, of photoreactive reagent, 46
Indirect reaction, 170, 171
In situ hybridization, 66, 94, 111, *see also* Hybridization
In vitro transcription, 48
Ionic strength, 125
Isotope, radioactive, 28

323

Index

K

Klenow
    enzyme, 22, 23, 24, 25, 26, 27
    polymerase, 22

L

Label(s)
    antigenic, 44, 138
    biotin, 171
    choice of, 164
    criteria for determining, 21
    enzymatic, 158, 175
    fluorescent, 158
    particular, 158
    radioactive, 137, 138
Labeling
    by addition, 46
    antigenic, 39
    artifactual, 147
    controls for, 56
    by 5′ extension, 44
    multiple, 159
    oligonucleotide, 40
    quality of, 167
    radioactive, 38
Leuckart bars, 83, 85
Levamisole, 176
Light green, 192
Light microscopy, 197, 200
Liquid emulsion, 146

M

Macroautoradiography, 145, 241
Matched hybrid, 123
Mercury-vapor lamp, 47, 204
Messenger RNA, 7
Methylene blue, 191
Methylene eosinate blue solution, 192
Methyl green, 191, 194
Microautoradiography, 145, 150, 153
Microscope(s)
    classification of, 199
    light, 197
Mismatched hybrid, 123
Modes of observation, 242
Moisture chamber, 132, 136, 171, 184
Monoclonal IgG, 169
Monosodium phosphate, 95
Morphology, preservation of, 88
Mounting
    after dehydration, 242
    in aqueous solution, 249
    medium, 194, 195, 250
    permanent, 249

N

Na^+ concentration, 122
NaOH, 95, 104
Neutral stains, 192
Nick translation, 8, 15, 22, 23, 24, 28, 29, 30, 31, 40, 56, 118
Nitroblue tetrazolium, 175
Non-sense probe, 10
Nuclear stains, 191
Nucleic acid, 7
    preservation of, 213
    structure of, 114
Nucleotide(s), 115
    antigenic, 29, 30
    labeled, 24, 53
    missing, 27
    modification of, 13
    radioactive, 44
    unlabeled, 29, 30

O

Observation, 213, 216, 242
Oligonucleotide(s), 7
    5′ extension of, 43
    labeled, 40, 45
    probe(s), 9, 121
        antigenic, 230, 244
        protocol for, 133
        radioactive, 226, 238
    synthetic, 20
Orceine, 191
Osmolarity, 67
Oxidation, risk of, 14
Oxido-reduction reaction, of NBT-BCIP system, 176

P

Palindromic sequences, 9
Paraffin
    sections
        obtaining, 86
        preparation of, 85
    wax, tissue embedded in, 81
Paraformaldehyde, 70, 75, 95
Particle label, 163
PCR, *see* Polymerase chain reaction
PEG, *see* Polyethylene glycol
Pepsin, 95, 100, 101
Perfusion, fixation by, 70, 71
Permeabilization, 98
Peroxidase, 158, 205, 248
Phase contrast light microscopy, 200
Phosphodiester bonds, 115, 116
Phosphorylation, 45
Photoactive antigen, 47
Photo-digoxigenin, 19
Photography
    emulsions, 145, 151

enlargement, 199
stability for, 198
Photoreactive reagent, incubation of, 46
Plasmid, 11
　DNA, 31
　insertion into, 49
　linearization of, 51
Poly A, 127
Polyclonal IgG, 169
Polyethylene glycol (PEG), 96
Polymerase chain reaction (PCR), 31
　amplification by, 31
　asymmetric, 36, 37, 39
　in situ, 218
　labeling by symmetrical, 33
　protocol, 35
Polymerization, 16, 68
Polynucleotides, synthetic, 22
Postfixation, 94, 100
Potassium cacodylate, 41
Precipitating fixatives, 67, 69
Precipitation
　DNA, 46
　ethanol, 26, 27, 54
　techniques, 54
Prehybridization, 94, 103, 215
Pretreatments, 89–106, 235
　acetylation, 102
　　aim, 102
　　protocol, 102
　dehydration, 102–103
　　aim, 102
　　protocol, 103
　denaturation of target nucleic acid, 104–106
　　aim, 104–105
　　protocols, 105–106
　deproteinization, 99–102
　　aim, 99
　　chemical treatment, 102
　　enzymatic treatment, 99–102
　equipment/reagents/solutions, 95–97
　　equipment, 95
　　reagents, 95–96
　　solutions, 96–97
　permeabilization, 98
　　agents, 98
　　aim, 98
　　protocols, 98
　prehybridization, 103–104
　　aim, 103
　　protocol, 103–104
　principle, 93
　stabilization of structures, 97–98
　　aim, 97
　　protocol, 97–98
　summary of different steps, 94
Primer(s), 22, 34
　anti-sense, 37
　application of, 33
　nonanucleotide, 26

Probe(s), 1–60, 93
　antigenic, 58, 130, 203, 234, 243
　anti-sense, 10
　attaching of itself to target nucleic acid sequence, 111
　cDNA
　　antigenic, 228
　　radioactive, 225
　concentration, 215
　controls for labeling, 56–58
　　antigenic probes, 58
　　radioactive probes, 56–57
　cRNA, 121
　　antigenic, 245
　　protocol for, 135
　　radioactive, 227, 239
　denaturation of, 60
　DNA, 59
　fragmentation of, 57
　labeling techniques, 22–53
　　addition, 46–48
　　amplification by asymmetric PCR, 36–39
　　amplification by PCR, 31–35
　　3′ extension, 40–43
　　5′ extension, 43–45
　　in vivo transcription, 48–53
　　nick translation, 27–31
　　random priming, 22–27
　labels, 12–21
　　antigenic labels, 17–20
　　criteria of choice, 21
　　radioactive labels, 13–16
　lability of, 49
　length of, 122
　non-sense, 10
　oligonucleotide, 121
　　antigenic, 230, 244
　　protocol for, 133
　　radioactive, 226, 238
　precipitation techniques, 54–56
　　drying, 56
　　ethanol precipitation, 54–55
　　passage through column, 55–56
　radioactive, 25, 42, 126, 203, 234, 236
　radiolysis of, 57
　RNA, 48, 121
　sense, 10, 217
　sequence, determination of, 10
　storage/use, 59–60
　　storage, 59
　　use, 59–60
　synthesis, 11, 36
　types of, 8–12
　　criteria of choice, 12
　　double-stranded DNA probes, 8
　　oligonucleotide probes, 9–10
　　RNA probes, 11
　　single-stranded DNA probes, 8–9
Problems, *see* Controls and problems
Pronase, 95, 100, 101
Proteinase, 97
Proteinase K, 95, 99

Proteins, 52, 124
Protocols, typical, 221–250
    antigenic probes, 243–246
        cDNA probe, 243–244
        cRNA probe, 245–246
        oligonucleotide probe, 244–245
    autoradiography, 241–242
        autoradiographic developing, 241
        counterstaining, 241
        macroautoradiography, 241
        microautoradiography, 241
        modes of observation, 242
        mounting after dehydration, 242
    immunocytological detection, 247–250
        counterstaining, 249
        immunohistological reaction, 247
        mounting, 249–250
        revelation with alkaline phosphatase, 247–248
        revelation with peroxidase, 248–249
    preparation of probes, 225–231
        antigenic cDNA probe, 228–229
        antigenic cRNA probe, 230–231
        antigenic oligonucleotide probe, 230
        radioactive cDNA probe, 225
        radioactive cRNA probe, 227–228
        radioactive oligonucleotide probe, 226
    preparation of tissue, 232–234
        frozen tissue, 232
        paraffin-embedded tissue, 233
        preparation of cells, 234
    pretreatments, 235–236
    radioactive probes, 236–240
        cDNA probe, 236–237
        cRNA probe, 239–240
        oligonucleotide probe, 238–239

R

Radiation, 144, 145
    energy, 146
    scale, 200
Radioactive probes, *see* Probes, radioactive
Radioisotopes, 16, 206
Radiolysis, 16
Radioprotection, 15, 16
Random priming, 22
Rapid nuclear red, 194
Reactive medium, preparation of, 237, 243, 246
Recapitulative table, 219–220
Refractive index, 195
Rehydration, 94, 233
Replication, 118
Revelation, 126, 137, 213, 216, 218
    alkaline phosphatase, 247
    peroxidase, 248
    solution, 248
Revelation of hybrids, 139–185
    antigenic hybrids, 156–185
        amplification by tyramide, 180–185

        detection, 167–173
        immunohistological reaction, 157–166
        revelation, 173–180
    choice of method, 143
    radioactive hybrids, 143–156
        characteristics, 145–147
        macroautoradiography, 147–150
        microautoradiography, 150–156
        principle, 144
Rhodol green, 160
Ribonucleic acid (RNA), 7, 95, 115, 117
    conservation of, 66
    degradation of, 53
    hybrids, 49
    messenger, 7
    probes, 11, 48, 121
    single-stranded, 117
    transfer, 7, 50, 119
    viral, 7
Ribonucleotide(s)
    radioactive, 50, 51
    triphosphate, 15
Ribose, 114
RNA, *see* Ribonucleic acid
RNase(s)
    A, 127
    activation of, 212
    -free reagents, 56
    treatment, 137

S

Saccharose, 76
Saline concentration, 121
Saponin, 95, 98
Sarcosyl, 127
Scintillation fluid, 57
Sense probe, 10, 11, 217
Sequence(s)
    comprising coding region, 9
    homology of, 114, 123
    palindromic, 9
    probe, 10
Signal
    /background ratio, 211
    /noise ratio, 104
    observation of, 205
Silica gel, 81, 95, 127
Single-stranded DNA, 7, 25, 213
Sodium chloride, 96, 127
Southern blot verification, 218
Stabilization, of structures, 97
Stain(s), *see also* specific types
    choice of, 192
    nuclear, 191
Sterilized water, 128
Steroids, 18
Storage methods, 59

Streptavidin, 17, 157
    conjugates, 166
    protocol, 172
Streptomyces avidinii, 17, 157
Stringency conditions, 124

T

Target nucleic acid, 93, 94
    denaturation of, 104
    sequence, 111
TdT, *see* Terminal deoxytransferase
Terminal deoxytransferase (TdT), 41, 42
Tetramethyl rhodamine isothiocyanate, 160, 161
Texas red sulfonyl chloride, 160, 162
Thymine, 116, 119
Tissue preparation, 61–88
    fixation, 66–73
        equipment/solutions, 70
        principle, 66–68
        protocols, 71–73
        types of fixative, 68–70
    frozen tissue, 74–81
        advantages/disadvantages, 81
        cryoprotection, 76
        equipment/reagents, solutions, 75
        freezing, 76–78
        preparations of frozen sections, 78–81
        summary of different steps, 74
    tissue embedded in paraffin wax, 81–88
        advantages/disadvantages, 88
        dewaxing of sections, 87–88
        embedding in paraffin wax, 84–85
        equipment/reagents/solutions, 83
        preparation of paraffin sections, 85–87
        summary of different steps, 82

tissue removal, 65–66
    precautions, 66
    tissue origin, 65
Toluidine blue, 154, 191, 193
Transfer RNA, 7, 50, 119
Triethanolamine, 96
Tris-hydroxymethyl-aminomethane, 96
Trisodium citrate, 96, 127
Tritium, 15
Triton X-100, 96, 98
Tyramide
    amplification by, 180
    –biotin formula, 181
    conjugates, 182

U

Uracil, 117, 119

V

Viral RNA, 7

W

Washing, 213, 216, 218
    curve, for hybridization, 125
    purpose of, 123
    time, 125
Water
    distilled, 154, 242
    sterilized, 128

X

Xylene, 88, 96, 153